This man, in words of Emerson's,
carries the holiday in his eye;
he is fit to stand the gaze of millions.

Harvard Film Studies

Pursuits of Happiness

THE HOLLYWOOD COMEDY OF REMARRIAGE

Stanley Cavell

Harvard University Press
Cambridge, Massachusetts, and London, England

Copyright © 1981 by the President and Fellows of Harvard College
All rights reserved
Printed in the United States of America
Tenth printing, 2003
Library of Congress Cataloging in Publication Data

Cavell, Stanley, 1926–
   Pursuits of happiness.

   Includes index.
   1. Comedy films—History and criticism.   2. Marriage
in moving-pictures.   I. Title.
PN1995.9C55C38   791.43′09′09354   81-6319
ISBN 0-674-73905-1 (cloth)   AACR2
ISBN 0-674-73906-X (paper)

To Benjamin William Cavell

CONTENTS

Pursuits of Happiness

INTRODUCTION

WORDS FOR A CONVERSATION

ACH of the seven chapters that follow contains an account of my experience of a film made in Hollywood between 1934 and 1949, an account guided by two claims. The first claim is that these seven films constitute a particular genre of Hollywood talkie, a genre I will call the comedy of remarriage. I am for myself satisfied that this group of films is the principal group of Hollywood comedies after the advent of sound and therewith one definitive achievement in the history of the art of film. But I will not attempt to argue directly for that here, any more than I will attempt explicitly to convince anyone that film is an art. The second guiding claim of these accounts is that the genre of remarriage is an inheritor of the preoccupations and discoveries of Shakespearean romantic comedy, especially as that work has been studied by, first among others, Northrop Frye. In his early "The Argument of Comedy," Frye follows a long tradition of critics in distinguishing between Old and New Comedy: while both, being forms of romantic comedy, show a young pair overcoming individual and social obstacles to their happiness, figured as a concluding marriage that achieves individual and social reconciliations, New Comedy stresses the young man's efforts to overcome obstacles posed by an older man (a senex figure) to his winning the young woman of his choice, whereas Old Comedy puts particular stress on the heroine, who may hold the key to the successful conclusion of the plot, who may be disguised as a boy, and who may undergo something like death and restoration. What I am calling the comedy of remarriage is, because of its emphasis on the heroine, more intimately related to Old Comedy than to New, but it is significantly different from either, indeed it seems to transgress an important feature of both, in casting as its heroine a married woman; and the

1

drive of its plot is not to get the central pair together, but to get them *back* together, together *again*. Hence the fact of marriage in it is subjected to the fact or the threat of divorce. A significant question for us is therefore bound to be: How is it that this transformation is called for when classical comedy moves to film?

I habitually call these accounts of films "readings" of them. What I mean by reading a film as well as what I conceive a genre of film to be (matters internal to what I think film is) will receive specification in the course of the discussions themselves. Films other than the ones I give readings of belong to the genre of remarriage comedy; six or seven of them are cited along the way. But I take the seven featured here to be definitive of the genre, the best of the genre, worthy successors of the great comedies of the Hollywood silent era. Worthier than the Marx brothers or W. C. Fields? I might answer this by distinguishing the comedy of clowns from the romantic comedy of manners. Or I might rather answer by saying that while the characters of the comedy of remarriage are not worthier or funnier or deeper than the characters projected by the Marx brothers and by Fields, and the individual actors not specifically as gifted for comedy, the films as films of the comedy of remarriage are worthier successors of the great films as films of Chaplin and Keaton. Such claims are at best staked out in the pages that follow; a test of them awaits their fate under the pressure of whatever counterclaims may be advanced against them.

All but one of the seven films centrally in question for me appear within the seven years from 1934 to 1941; hence they, and other films to be distinguished from them, are often referred to as Hollywood thirties comedies. Why they emerge and disappear over the years in question are matters our discussions ought to provide terms for understanding. The explanation I have heard for this historical phenomenon—and it seems to have become something of a piece of folk wisdom—is that thirties comedies were fairy tales for the Depression. This can hardly be denied if what it means is that in a time of economic depression romances were made in Hollywood that took settings of immense luxury and that depicted people whose actions often concerned the disposition of fantastic sums of money. If luxurious settings and fantastic sums of money were confined to the Hollywood films of this period, and if Hollywood films of luxury and expenditure were confined to works that fit the genre of remarriage, then I would be more drawn to an economic interpretation of the films I have interested myself in, or to an explanation of the emergence of the genre by economic causation. Since the facts are otherwise it matters to me that that explanation does not specifically account for the form in question.

There are comedies of the period which might better fit the description "fairy tales for the Depression," ones like *If I Had a Million* (1933), which consists of a set of episodes about what happens to various people when at random they are handed the title sum of money. But this seems less a reflection of particular economic realities or fantasies than of the ancient theme of fairy tales concerning the unforeseeable consequences of having wishes granted, call this the fantasy of escaping the realm of economy altogether.

Or take the more famous *You Can't Take It with You* (Frank Capra, 1938). An honest but poor young girl (Jean Arthur) and the son (James Stewart) of a rich father (Edward Arnold) are in love and want to marry; unknown to them the girl's beloved grandfather (Lionel Barrymore) is all that stands in the way of the boy's father's scheme to buy up the houses of all the girl's friends and neighbors and throw them out to make way for a munitions factory that is the key to the biggest deal in contemporary business. The grandfather will not sell his house and without it the factory somehow will not fit into the remainder of the twelve square blocks the financier has bought up. Grandpa won't sell for various reasons. One is that he knows he and his house are all that can prevent his entire neighborhood from destruction. Another reason is that his granddaughter's father and two friends of his spend all their time in the basement of the house inventing and making things; importantly, making munitions, I mean fireworks, just for the fun of it, for which the local police take them to be Communists. A third reason is that he had been happily married in this house, and while his wife died before his granddaughter was born, nevertheless the wife's presence, even her sweet odor, remains in the house, concentrated as it happens in the room that his granddaughter occupies.

The reasons not to sell go dead when the girl disappears, unable to tolerate the differences between her and the boy's families. Grandpa almost instantly sells the house to the boy's mean father and plans to move to make a new home for the granddaughter, away from what makes her unhappy. The image of this house of romance, of whim and acceptance fulfilled every day, as Emerson promised us, near the end stripped of its life and ready for removal, is meant I guess to strike us with the force of the end of *The Cherry Orchard*. But where is the inevitability? Grandpa *can* take it with him, I mean take the money from the sale and buy a new house; but why *must* he? What is supposed to make it credible that this putatively good old man, urging everybody to do what he or she likes, to have the courage of his or her happiness, an Emersonian sage, is willing on an instant's notice to leave his entire neighborhood to destruction because he has to follow his grown granddaughter who is having trouble with her boyfriend? Is this an expression of the courage for happiness? Or is it proof that blood is thicker than water? Some Emersonian sentiment. Or are we to realize that Grandpa, exactly because he is the only neighbor who privately owns his own place, is the one whose solidarity with his neighbors is mostly talk, and that in the end he is closer to the mean big people than he is to the good little people? Surely Capra, whatever his problems with endings, could have avoided so naked a revelation and conflict of values if he had wanted to. For example, he could have saved the house (for the neighbors) the way he saved the house and family some years later in *It's a Wonderful Life* (1946), by taking up a collection from the little people; or the young man could secretly have raised the money, and when the girl finds this out she returns and . . .

Evidently we need a more credible explanation of Grandpa's motivation. He follows the girl not because she cannot recover without him but because he cannot live without her. (He may not have been prepared to sell the house to the young man.) She is the sweetness of his life. When sweetness and social solidarity conflict there may be tragedy, and in this world they will conflict. Besides, Grandpa is not proven wrong in the event. News of his plans brings the girl back in order to stop him; and his actions help make the boy's father relent, which means help him find the courage to do as he likes, which is not to make munitions (that only upsets his stomach) but to play the harmonica. Without offering this as a general solution to the problem of arms limitation, I hope it may allow us to see the value of this film not as a study in neighborhood organizing but as a vision of community, Utopian no doubt. The meaning of the vision is not so much that organization requires hope, which requires vision, as it is that happiness is not to be won just by opposing those in power but only, beyond that, by educating them, or their successors. Put otherwise, the achievement of human happiness requires not the perennial and fuller satisfaction of our needs as they stand but the examination and transformation of those needs.

Even if one whole-heartedly agreed with such a thought (as voiced, say, in Plato and in Rousseau and in Thoreau and in Freud) no one would say that it is applicable in all human contexts. It applies only in contexts in which there is satisfaction enough, in which something like luxury and leisure, something beyond the bare necessities, is an issue. This is why our films must on the whole take settings of unmistakable wealth; the people in them have the leisure to talk about human happiness, hence the time to deprive themselves of it unnecessarily. Emerson, while we are at it, in his essay "History," has expressed the best way I know of initially understanding these settings: "It is remarkable that involuntarily we always read as superior beings . . . We honor the rich because they have externally the freedom, power, and grace which we feel to be proper to man, proper to us. So all that is said of the wise man by Stoic or Oriental or modern essayist, describes his unattained but attainable self."

But when I spoke a moment ago of the depicting of the disposition of fantastic sums of money I did not mean that the sums had necessarily to be large but that large or small the amounts had to be significant. In *It Happened One Night* Clark Gable is not interested in a $10,000 reward but he insists on beng reimbursed in the amount of $39.60, his figure fully itemized. The economic issues in these films, with all their ambivalence and irresolution, are invariably tropes for spiritual issues. (Which is not to deny that they can be interpreted the other way around too, the spiritual conflicts as tropes for the economic. These conflicts are bound up with the conflict over the direction of interpretation, the question, say, of what money, and how you get it, can make you do.) This is what we might expect of American romantic, or Utopian, works. The figure Gable claims is owed to him is of the same order as the figure, arrived at with similar itemization, Thoreau claimed to have spent in building his house, $28.12½. The purpose of these men in both cases is to distinguish themselves, with poker faces, from those who do not know what things cost, what life costs, who do not know what counts. It is as essential for the settings of our films to be such that we can expect the characters in them to take the time, and take the pains, to converse intelligently and playfully about themselves and about one another as it is essential for the settings and characters of classical tragedy to be such that we can expect high poetry from them. Our critical task is to discover why they use their time as they do, why they say the things they say. Without taking up the details of the films we should not expect to know what they are, to know what causes them.

I am assuming that the films may themselves be up to reflecting on what it is that causes them, hence that they may have some bearing, for instance, on our experience and understanding of the Depression. *It Happened One Night* is a film, I will come to say, about being hungry, or hungering, where hungering is a metaphor for imagining, in particular imagining a better, or satisfying, way to live. There are a number of foods in the film, forming a little system of symbolic significance. There is also a woman, in what I call a "Depression vignette," who faints from hunger. What is the relation of the symbolism to this vignette? Has Capra stuck in the vignette to buy off criticism of his treating of the problems of leisure in an age of desperation? Or as a confession that he has no solution to give us to the problem of hunger and so might be excused for providing some distraction from it, which he does have to give us? Or is he really to be understood as taking the occasion of the Depression to ask what it is we as a people are truly depressed by, what hunger it is from which we all are faint? And if he is to be understood so, isn't this worse, morally speaking, than making up fairy tales? Wouldn't it be aestheticizing human suffering, or transcendentalizing it—like saying "Man does not live by bread alone" to a man in a breadline?

But then this is a risk any serious art must run that opens itself to present suffering, a risk run by, say, the famously beautiful prose and photographs of James Agee and Walker Evans in their *Let Us Now Praise Famous Men* as well as a hundred years earlier by Emerson in speaking of those living in "silent melancholy" and by Thoreau in describing the mass of men as leading "lives of quiet desperation." Does one conceive that Emerson and Thoreau are writing for someone other than the ones they describe out of their perception of the nation's depression? Mostly there is no one else. Or does one conceive that the despair they perceive is essentially a spiritual one, the kind a transcendentalist can see, and therefore betokens not so serious a hunger? They knew the accusation of refusing to help those whom they saw in need, as if giving what they wrote were less practical than alms, and they answer the accusation openly. Around the middle of *Walden* Thoreau shows himself offended by the impoverished, inefficient lives of a certain John Field and his family and berates them for not reckoning cost as he does. I do not know that this passage takes upon itself a greater hardness, though the hardness is given greater specificity, than Emerson's saying in "Self-Reliance," as he pictures himself going off to write, "Do not tell me, as a good man did to-day, of my obligation to put all poor men in good situations. Are they *my* poor?" That is, it is not I who make them and who keep them poor; and so far as I can better the situation of whoever is poor I can do it only by answering my genius when it calls. But to give that sort of answer one must have a healthy respect for the value of one's work, let us say for its powers of instruction and redemption.

Is it obvious that the makers of the films we will read through— Frank Capra, Leo McCarey, Howard Hawks, George Cukor, Preston Sturges—are in principle not entitled to such claims for their work? Would the principle be that film cannot provide such instruction, or that American films cannot, or that Hollywood comedies (at least those after the silent period) cannot? Why should one believe any of this? Of course these films *can* be appropriated by any or all of their fans as fairy tales rather than, let us say, as spiritual parables. But so can Scripture be similarly appropriated; so can Emerson and Thoreau; so can Marx and Nietzsche and Freud. But from what better writers can one learn, or have companionship in knowing, that to take an interest in an object is to take an interest in one's experience of the object, so that to examine and defend my interest in these films is to examine and defend my interest in my own experience, in the moments and passages of my life I have spent with them. This in turn means, for me, defending the process of criticism, so far as criticism is thought of, as I think of it, as a natural extension of conversation. (And I think of conversation as something within which that remark about conversation is naturally in place. This one too.) I will do some of this defending once it begins to emerge that these films are themselves investigations of (parts of a conversation about) ideas of conversation, and investigations of what it is to have an interest in your own experience.

There will be resistance to considering the films in the way I do beyond the appropriating of them as escapist material for a particular period. Before moving from the concept of the Depression I note that

Malcolm Cowley, sifting his attentive experience of the period and of its writing, picks out three features for emphasis that our films may be seen to share.* The transcendentalist possibility I was noticing seems to be what Cowley calls the period's millennialism, as if under the depression an ecstasy were discernible; he also mentions the presence in a number of the period's good novels of the theme of death and rebirth; and he finds a chorus of witnesses to the dignity of man. As we progress these themes will be found to play curiously sensitive roles in our set of films. But to see this we will have to develop a certain skepticism about appearances. For example, it will be a virtue of our heroes to be willing to suffer a certain indignity, as if what stands in the way of change, psychologically speaking, is a false dignity; or, socially speaking, as if the dignity of one part of society is the cause of the opposite part's indignity, a sure sign of a disordered state of affairs.

I AM NOT INSENSIBLE, whatever defenses I may deploy, of an avenue of outrageousness in considering Hollywood films in the light, from time to time, of major works of thought. My sense of the offense this can give came to a climax in presenting a draft of my essay on It Happened One Night (Chapter 2) to a university symposium entitled "Intellect and Imagination: The Limits and Presuppositions of Intellectual Inquiry." This essay begins with the longest consecutive piece of philosophical exposition in the book, concerning the thought of Immanuel Kant, whose teaching has claim to be regarded as the most serious philosophical achievement of the modern age. And what follows this beginning is the discussion of a Frank Capra film, not even something cinematically high-minded, something sad and boring, something foreign or foreign-looking, or something silent. Evidently I meant my contribution to a discussion of limits and their transgressions to be an essay that itself embodies a little transgression in its indecorous juxtaposition of subjects. I introduced my discussion of that essay at the symposium by giving three reasons for my transgression, that is, for courting and expressing a certain outrage.

First, I wished to take the opportunity to acknowledge that philosophy, as I understand it, is indeed outrageous, inherently so. It seeks to

* *And I Worked at the Writer's Trade* (New York: Viking Press, 1978).

disquiet the foundations of our lives and to offer us in recompense nothing better than itself—and this on the basis of no expert knowledge, of nothing closed to the ordinary human being, once, that is to say, that being lets himself or herself be informed by the process and the ambition of philosophy. Wittgenstein voices the accusation against his work that it "seems to destroy everything interesting, that is, all that is great and important." He replies, as translated, that what he is "destroying is nothing but houses of cards"—as if this destruction were less important, less devastating than some other, as if we had any other modes of dwelling.*

Second, I wished to take the occasion of a symposium to raise a question of the limits of the convival, anyway of the extent to which the experiences and the pleasures of the participants were sharable—a way of testing the limits or the density of what we may call our common cultural inheritance. This issue was focused for me by the request of several participants for a thumbnail sketch of Kant's views against which one unfamiliar with Kant might assess my claims about him in my opening pages. (And assess echoes in the closing?) Since my pages on Kant are already a thumbnail sketch, I assumed that what was being requested was a preceding sketch, maybe like a short encyclopedia entry. Whatever the value of such a genre, for my purposes it would have none. It would not, for example, put its recipient in a position to assess certain originalities in the way I sketched Kant's vision. A purpose of mine, in any case, was precisely to bring into question the issue of our common cultural inheritance. The request for a (another) thumbnail sketch is an expression of something my sketch, in its juxtaposition with a Hollywood film, itself registers, that Kant is not a part of the common cultural inheritance of American intellectuals. (Perhaps this just means that we are not Germans or Central Europeans.) But if one of the indisputably most important philosophical achievements of the modern era of Western civilization is not a piece of our inheritance, what is? The ensuing discussion of a Hollywood film might stand in the place of an answer, or as a certain emblem of an answer. It must be an ambiguous place. One ought not to say, for example, that we have films instead of books as our legacy. In the first place, we do have books; in

---

* I respond a little differently to Wittgenstein's observation in my Foreword to *The Claim of Reason* (New York: Oxford University Press, 1979).

the second, it is not clear that we do have films in common, or not clear what it is to "have" them; in the third, the idea of "instead of" is undefined. The fact is that you cannot acquire the Kant I know from me, certainly not here and now. Anyway, would this work be worthwhile just for the sake of having something intellectual in common? Whereas a companion fact is that you can acquire from me, or reacquire, a Hollywood film, here and now (if you've seen it recently), along with certain related matters. But would this be something worth having in common?

My juxtaposition of Kant and Capra is meant to suggest that you cannot know the answer to the question of worthwhileness in advance of your own experience, not the worthwhileness of Capra *and* not that of Kant. (Some might feel this means that nothing we stand to inherit is sacred, and further that this just means we are Americans.) I am not, in the case of the Capra, simply counting on our capacity for bringing our wild intelligence to bear on just about anything, say our capacities for exploring or improvisation. What we are to see is the intelligence that a film has *already* brought to bear in its making; and hence perhaps we will think about what improvisation is and about what importance is.

Perhaps we will not, too; which means that my transgressing conjunction of interests will be refused as a courting, and an expression, of the outrageous. This would tend to outrage me (because it would strike me as intellectually complacent and neglectful)—to acknowledge which is the third reason for my conjunction of film and philosophy.

To subject these enterprises and their conjunction to our experience of them—that is, to assess our relation to these enterprises—is a conceptual as much as an experiential undertaking; it is a commitment to being guided by our experience but not dictated to by it. I think of this as checking one's experience. I indicated a moment ago by my quotation from Wittgenstein that philosophy requires the sense of the title of all that is great and important to be given up to experience. If one may think of this as an overcoming of philosophical theory, I should like to stress that the way to overcome theory correctly, philosophically, is to let the object or the work of your interest teach you how to consider it. I would not object to calling this a piece of theoretical advice, as long as it is also called a piece of practical advice. Philosophers will naturally assume that it is one thing, and quite clear how, to let a philosophical

work teach you how to consider it, and another thing, and quite obscure how or why, to let a film teach you this. I believe these are not
such different things.

A READING OF A FILM sets up a continuous appeal to the experience of the
film, or rather to an active memory of the experience (or an active anticipation of acquiring the experience). It seems to me that even those who
are willing to believe that the details of every motion and position of
what the camera depicts, and of every motion and position of the camera that is doing the depicting, may be significant in determining what a
film is about—to believe, that is, that the visual facts of a movie you
care about may survive the same kind of attention you would give the
verbal facts of a literary text you care about—even among these people
it is hard to believe that the *words* spoken in the film should be taken
with the same seriousness. It is true that the words of dialogue put on
the page seem too poor to carry the significance I will attach to them.
And in a sense this is right—they have to be taken from the page and
put back, in memory, onto the screen. It is natural to neglect this obligation because words can be quoted on the page and moving images
cannot be, so you can think that work has been done for you (by the
words on the page) when the work for you to do has only been con-
veniently notated. Apart from a clear recall, or a vivid imagination, of
these words as spoken by these actors in these environments, my attention to the words may well seem, indeed ought to seem, misplaced
or overdone. (Something analogous is familiar in reading plays. Even
Ibsen's words might seem too poor on the page to live up to their reputation. Let this indicate, without denying that film is a visual medium,
that film is a medium of drama.) This is an epitome of the nature of
conversation about film generally, that those who are experiencing
again, and expressing, moments of a film are at any time apt to become
incomprehensible (in some specific mode, perhaps enthusiastic to the
point of folly) to those who are not experiencing them (again). I am re-
garding the necessity of this risk in conversing about film as revelatory
of the conversation within film—at any rate, within the kind of film
under attention here—that words that on one viewing pass, and are
meant to pass, without notice, as unnoticeably trivial, on another reso-
nate and declare their implication in a network of significance. These

film words thus declare their mimesis of ordinary words, words in daily conversation. A mastery of film writing and film making accordingly requires, for such films, a mastery of this mode of mimesis.*

Checking one's experience is a rubric an American, or a spiritual American, might give to the empiricism practiced by Emerson and by Thoreau. I mean the rubric to capture the sense at the same time of consulting one's experience and of subjecting it to examination, and beyond these, of momentarily *stopping*, turning yourself away from whatever your preoccupation and turning your experience away from its expected, habitual track, to find itself, its own track: coming to attention. The moral of this practice is to educate your experience sufficiently so that it is worthy of trust. The philosophical catch would then be that the education cannot be achieved in advance of the trusting. Hence Emerson is logically forced to give his best to Whim. Yet the American inheritance of Kant (and wasn't this in advance of experience?) is essential to making up Transcendentalism, and hence it goes into what makes Emerson Emerson and what makes Thoreau Thoreau. Encouraged by them, one learns that without this trust in one's experience, expressed as a willingness to find words for it, without thus taking an interest in it, one is without authority in one's own experience. (In a similar mood, in *The Claim of Reason*, I speak of being without a voice in one's own history.) I think of this authority as the right to take an interest in your own experience. I suppose the primary good of a teacher is to prompt his or her students to find their way to that authority; without it, rote is fate. The world, under minimum conditions of civilization, could not without our cooperation so thoroughly contrive to separate us from this authority. Think of it as learning neither to impose your experience on the world nor to have it imposed upon by the world. (These are sorts of distortions of reason Kant calls fanaticism and superstition.) It is learning freedom of consciousness, which you might see as becoming civilized. Unless spoken from such a position, why should assertions concerning the value of, for example, film be of any concern to us?

---

* I claim in "Ending the Waiting Game: A Reading of *Endgame*," in *Must We Mean What We Say?* (Cambridge, England: Cambridge University Press, 1976), pp. 128–30, that Beckett achieves a new way for theater of accomplishing this point of mimesis. A reliable transcript of the dialogue of *It Happened One Night*, together with, instructively, a pervasively inaccurate set of descriptions and "stage directions," is in *Four-Star Scripts*, ed. Lorraine Noble (New York: Doubleday, Doran and Co., 1936). Another published script of a principal remarriage comedy is *Adam's Rib* (New York: Viking Press, 1972).

It is fundamental to this view of experience not to accept any given experience as final but to subject the experience and its object to the test of one another. For this a concept such as that of, let me say, the good encounter must come into play. There are such things as inspired times of reading or listening as surely as there are such things as inspired times of writing or composition. Successive encounters of a work are not necessarily cumulative; a later one may overturn earlier ones or may be empty. A valuable critic tends to know of his or her experience which is which as surely as he knows about an object what is what. A work one cares about is not so much something one has read as something one is a reader of; connection with it goes on, as with any relation one cares about. (Thoreau's copy of Homer is open on his table at Walden. So far as philosophy is a matter of caring about texts, meditation is its work before argumentation, since the end of the caring cannot be expressed in a conclusion which you might *take away* from the text.) Yet everything in our film culture, and not only there, has until recently conspired to adopt as standard the experience taken on one viewing. My impression is that most people still see all films except certain private or cult obsessions just once, and reviewers review on one viewing, saying things that there will probably be no practical way to test. In each other art it is comparatively normal to expect to be able to *go back* to a work you care about, at least in reproduction. Revival houses, university programs of film studies, television's unending dependence on Hollywood past, and perhaps any minute now video discs and cassettes, are changing these expectations. If these changes in mere practicality reach the point of making the history of film as much a part of the present experience of film as the history of the other arts is part of their present, this will result in a greater alteration of our experience of film, I predict, than any development since the establishment of the motion picture.

I SHOULD CONFESS that my confession to having courted a certain outrageousness in juxtaposing philosophy and film is not yet full, for I harbor the conviction that facing them with one another is positively called for; it is internal to my interest in each of them. From the side of film I have indicated in previous writings ways in which, as I might put it, film exists in a state of philosophy: it is inherently self-reflexive, takes itself as an inevitable part of its craving for speculation;* one of its seminal genres—the one in question in the present book—demands the portrayal of philosophical conversation, hence undertakes to portray one of the causes of philosophical dispute. It may be felt that these properties apply, more or less, to all the major arts. In that case what I am showing is that philosophy is to be understood, however else, aesthetically.

From the side of philosophy I can suggest what I see as its affinity for film by citing another passage of Emerson's, this time from "The American Scholar":

> I ask not for the great, the remote, the romantic; what is doing in Italy or Arabia; what is Greek art, or Provencal minstrelsy; I embrace the common, I explore and sit at the feet of the familiar, the low. Give me insight into today, and you may have the antique and future worlds. What would we really know the meaning of? The meal in the firkin; the milk in the pan; the ballad in the street; the news of the boat; the glance of the eye; the form and the gait of the body;—show me the ultimate reason of these matters; show me the sublime presence of the highest spiritual cause lurking, as always it does lurk, in these suburbs and extremeties of nature; . . . —and the world lies no longer a dull miscellany and lumber-room, but has form and order; there is no trifle, there is no puzzle, but one design unites and animates the farthest pinnacle and the lowest trench.

Something Emerson means by the common, the familiar, and the low is something I have meant (claiming the inheritance of the common preoccupation of J. L. Austin and of Wittgenstein), in my various defenses over the years of proceeding in philosophy from ordinary language, from words of everyday life. By "sitting at the feet" of the familiar and the low, this student of Eastern philosophy must mean that he takes the familiar and the low as his study, his guide, his guru; as much his point of arrival as of departure. In this he joins his thinking with the new poetry and art of his times, whose topics he characterizes as "the literature of the poor, the feelings of the child, the philosophy of the street, the meaning of household life." I note that when he describes himself as

* This is the theme of film's acknowledgment (or definition) of its medium, a preoccupation of *The World Viewed* and of "More of *The World Viewed*" as well as of the Foreword written for their joint reissue as *The World Viewed, Enlarged Edition* (Cambridge: Harvard University Press, 1979). The question of acknowledgment, of self-reflection, is not exhausted, as appears sometimes to be thought, by the tendency of films to be self-referential. The latter is at best a specialized (generally comic) mode of the former.

asking "not for the great, the remote, the romantic," he is apparently not considering that the emphasis on the low and the near is exactly the opposite face of the romantic, the continued search for a new intimacy in the self's relation to its world. His list of the matters whose "ultimate reason" he demands of students to know—"The meal in the firkin; the milk in the pan; the ballad in the street; the news of the boat; the glance of the eye; the form and gait of the body"—is a list epitomizing what we may call the physiognomy of the ordinary, a form of what Kierkegaard calls the perception of the sublime in the everyday. It is a list, made three or four years before Daguerre would exhibit his copper plates in Paris, epitomizing the obsessions of photography. I once remarked that Baudelaire, in his praise of a painter of everyday life, had had a kind of premonition of film.* Here I should like to add that without the mode of perception inspired in Emerson (and Thoreau) by the everyday, the near, the low, the familiar, one is bound to be blind to some of the best poetry of film, to a sublimity in it. Naturally I should like to say that this would at the same time ensure deafness to some of the best poetry of philosophy—not now its mythological flights nor its beauty or purity of argumentation, but now its power of exemplification, the world in a piece of wax.** It is to the point that the genre of film in question in the present book will at the end become characterizable as a comedy of dailiness.

In subjecting these films to the same burden of interpretation that I expect any text to carry that I value as highly, I am aware that there are those for whom such an enterprise must from the start appear misguided, those who are satisfied that they know what film is, that it is, for example, a commodity like any other, or a visual medium of popular entertainment (as compared with what?). But anti-intellectualism is no more or less attractive here than elsewhere. Neither, no doubt, is overintellectuality. If anti-intellectualism were the genuine corrective to overintellectuality then there would be no distinction between a sage and a punk. I am moved here to reiterate to the reader the sentiment I was expressing in speaking about the issue of a common cultural inher-

* *The World Viewed*, p. 42.
** Exemplification is a principal theme of *The Claim of Reason*. In "An Emerson Mood," included in an expanded edition of *The Senses of Walden* (forthcoming from North Point Press in Berkeley), I have spelled out a little further the idea of Emerson and Thoreau as underwriting the procedures and certain aspirations of Austin and Wittgenstein.

itance. This book is primarily devoted to the reading of seven films. If my citings of philosophical texts along the way hinder more than they help you, skip them. If they are as useful as I take them to be they will find a further chance with you.

THE THIRTIES were more than the Depression. They were phases of histories other than that of what is called the economies of nations. The opening years of the Depression were also the opening years of a new phase in the history of cinema, the years of the advent of sound. The year of the earliest member of our genre, 1934, is early enough for that film to have had a decisive say in determining the creation of Hollywood sound film. The genre it projected, on my interpretation, can be said to require the creation of a new woman, or the new creation of a woman, something I describe as a new creation of the human. If the genre is as definitive of sound comedy as I take it to be, and if the feature of the creation of the woman is as definitive of the genre as I take it to be, then this phase of the history of cinema is bound up with a phase in the history of the consciousness of women. You might even say that these phases of these histories are part of the creation of one another.

It may prove to be, at any rate, that this genre of film is in fact the main reason for positing the existence of such a phase in the consciousness, or unconsciousness, of women. This would be the case so long as the picture of the trajectory of the feminist movement looks the way it has been presented more than once in my hearing: that after the great figures and notable gains of the generations of women beginning with the Seneca Falls Convention in 1848 and culminating, so it seemed, in the winning of the vote for women in 1920, feminist thought and feminist practice somehow scattered themselves or lost their specific identity. After a decade to assess the value of suffrage there came the Depression, then the War, then the postwar Eisenhower generation of silence, then the civil rights movement for blacks, and only then, toward the end of the sixties, did a new phase of feminist history begin. As if the feminist preoccupation could not, during the four decades from the thirties through most of the sixties, get itself on the agenda of an otherwise preoccupied nation. I take the very existence of the genre of the comedy of remarriage—of course, on my interpretation of what its films are and what they are about—as proof that such a picture cannot be right. Coming from me, this claim is meant to be less about feminist theory and practice, about which my knowledge has barely begun, than about film, about the fact that films of the magnitude I claim the films in question in this book to be are primary data for what I would like to call the inner agenda of a culture. (I find Alice S. Rossi's description, in one of her introductions to a section of selections in *The Feminist Papers,* * closer to the view I am expressing: "The generation that followed the activist generation of suffragists may have been consolidating feminist ideas into the private stuff of their lives and seeking new outlets for the expression of the values that prompted their mothers' public behavior" [p. 616]. What I am saying differs in two ways from this sort of account. First, I am saying that there is no "may have been" about it, as if we needed better evidence. What I am looking for is the better interpretation of documents as blatant as, say, a constitutional amendment. Second, the idea of "the private stuff of their lives" is part of the intuition I wished to capture by speaking of an "inner agenda of a culture"; but beyond that I meant it to express the idea of something shared, call it a shared fantasy, apart from which the films under investigation here could not have reached their *public* position.)

The formulation "consciousness of women" is studiously ambiguous as between meaning the consciousness held with respect to women, whether by women or by men; and the consciousness held by women, with respect to themselves and everything else. By the consciousness of women as expressed in the genre of remarriage I mean something of both sides—I mean a development in the consciousness women hold of themselves as this is developed in its relation to the consciousness men hold of them. Whether in a given historical period and class and place this consciousness is fundamentally imposed upon women or whether the relation is one in which women are fundamentally equal partners in the development is something I assume it is the burden of history to show (the burden of its working and the burden of the students of its working). Our films may be understood as parables of a phase of the development of consciousness at which the struggle is for the reciprocity or equality of consciousness between a woman and a man, a study of the conditions under which this fight for recognition (as Hegel put it) or demand for acknowledgment (as I have put it) is a struggle for mu-

---

* (New York: Bantam Books, 1974).

tual freedom, especially of the views each holds of the other. This gives the films of our genre a Utopian cast. They harbor a vision which they know cannot fully be domesticated, inhabited, in the world we know. They are romances. Showing us our fantasies, they express the inner agenda of a nation that conceives Utopian longings and commitments for itself.

What suits the women in them—Claudette Colbert, Irene Dunne, Katharine Hepburn, Rosalind Russell, Barbara Stanwyck—for their leading roles? All were born between 1904 and 1911, about the years you would expect, given two assumptions: that the leading women must be around thirty years old as the genre is forming itself, neither young nor old, experienced yet still hopeful; and that within four or five years of the establishment of the talkie's material basis, it found in the genre of remarriage one of its definitive forms, as though cinema could barely wait to enter into the kind of conversation required of the genre and made possible by sound. An immediate significance of the women's being born in the latter half of the first decade of the century is that their mothers would have been of the generation of 1880, the generation of, for example, Eleanor Roosevelt, Frances Perkins, Margaret Sanger. A distinguished generation, one would think, and one is asked to think about it because in the fiction of our films the woman's mother is conspicuously and problematically absent. If these films are what I have called investigations of something like the creation of the woman in them, we are bound to ask what the absence of the maternal half of her creation betokens.

What is it about the conversation of just these films that makes it so perfectly satisfy the appetite of talking pictures? Granted the fact, the question can only be answered by consulting the films. Evidently their conversation is the verbal medium in which, for example, questions of human creation and the absence of mothers and the battle between men and women for recognition of one another, and whatever matters turn out to entail these, are given expression. So it is not sufficient that, say, the conversation be sexually charged. If it were sufficient then the genre would begin in 1931, with Noel Coward's *Private Lives*, a work patently depicting the divorce and remarrying of a rich and sophisticated pair who speak intelligently and who infuriate and appreciate one another more than anyone else. But their witty, sentimental, violent exchanges get nowhere; their makings up never add up to forgiving one another (no place they arrive at is home to them); and they have come from nowhere (their constant reminiscences never add up to a past they can admit together). They are forever stuck in an orbit around the foci of desire and contempt. This is a fairly familiar perception of what marriage is. The conversation of what I call the genre of remarriage is, judging from the films I take to define it, of a sort that leads to acknowledgment; to the reconciliation of a genuine forgiveness; a reconciliation so profound as to require the metamorphosis of death and revival, the achievement of a new perspective on existence; a perspective that presents itself as a place, one removed from the city of confusion and divorce. One moral to draw from the structure of *Private Lives* is that no one feature of the genre is sufficient for membership in the genre, not even the title feature of remarriage itself. Another moral is that the fact that *Private Lives* seems closer than our comedies do to the spirit of Restoration comedy is a good reason not to look to Restoration comedy (as I have periodically, for obvious reasons, found myself tempted to do) as a central source of the comedy of remarriage.

I FIND A PRECEDENT for the structure of remarriage, as said, in Shakespearean romance, and centrally in *The Winter's Tale*. This was one of the earliest and, while encouraging, most puzzling discoveries I made as I became involved in thoughts about the set of films in question here. Two puzzles immediately presented themselves. First, since Shakespearean romantic comedy did not remain a viable form of comedy for the English stage, compared with a Jonsonian comedy of manners, what is it about film that makes its occurrence there viable? This goes into the question why it was only in 1934, and in America of all places, that the Shakespearean structure surfaced again, if not quite on the stage. I have in effect already outlined the answer I have to that question. Nineteen thirty-four—half a dozen years after the advent of sound—was about the earliest date by which the sound film could reasonably be expected to have found itself artistically. And it happens that at that same date there was a group of women of an age and a temperament to make possible the definitive realization of the genre that answered the Shakespearean description, a date at which a phase of human history, namely, a phase of feminism, and requirements of a genre inheriting a remarriage structure from Shakespeare, and the nature of film's transformation of its human subjects, met together on the issue of the new creation of a woman. No doubt this meeting of interests is part of America's special involvement in film, from the talent drawn to Hollywood in making them to the participation of society as a whole in viewing them, and especially America's preeminence in film comedy.

The second puzzle about the Shakespearean precedent is why the film comedies of remarriage took as their Shakespearean equivalent, so to speak, the topic of divorce, which raises in a particular form the question of the legitimacy of marriage. Since I am saying that the comedy of remarriage does not look upon marriage as does either French farce or Restoration comedy, I had thought in vain about a comedic precedent for the remarriage form more specific than the Shakespearean. It finally dawned on me that the precedent need not be found in the history of comedy but in any genre to which the film comedies in question can be shown to have an exact conceptual relation. This thought permitted me to find an instance of what I was looking for in the most obvious place in the world I know of drama, in Ibsen, and particularly, it turns out, in *A Doll House*. (I learn to call it this, without the possessive, from a convincing explanation with which Rolf Fjelde prefaces his translation of the play.)* This is the latest of the ideas I introduce in these pages, and to commemorate it, and for future reference, I inscribe this early moment of my book with excerpts from the last pages of that play.**

NORA: Thank you for your forgiveness. (*She goes out through the door, right.*)

HELMER: No, don't go—What are you doing there?

NORA (*offstage*): Taking off my fancy dress.

HELMER: Yes, do that. Try to calm yourself and get your balance again, my frightened little songbird. Don't be afraid. I have broad wings to shield you. How lovely and peaceful this little home of ours is, Nora. . . . What's this? Not in bed? Have you changed?

NORA (*in her everyday dress*): Yes, Torvald. I've changed.

. . .

NORA (*after a short silence*): Doesn't anything strike you about the way we're sitting here?

* *Ibsen: The Complete Major Prose Plays* (New York: New American Library, 1978).
** Translated by Michael Meyer in *Ghosts and Three Other Plays* (New York: Anchor Books Original, 1966).

WORDS FOR A CONVERSATION

HELMER: What?

NORA: We've been married for eight years. Does it occur to you that this is the first time that we two, you and I, man and wife, have ever had a serious talk together?

HELMER: Serious? What do you mean, serious?

. . .

HELMER: Nora, how can you be so unreasonable and ungrateful? Haven't you been happy here?

NORA: No; never. I used to think I was; but I haven't ever been happy.

HELMER: Not—not happy?

NORA: No. I've just had fun. You've always been very kind to me. But our home has never been anything but a playroom. I've been your doll-wife, just as I used to be Papa's doll-child. And the children have been my dolls . . .

HELMER: There may be a little truth in what you say, though you exaggerate and romanticize. But from now on it'll be different. Playtime is over. Now the time has come for education.

NORA: Whose education? Mine or the children's?

HELMER: Both yours and the children's, my dearest Nora.

NORA: Oh, Torvald, you're not the man to educate me into being the right wife for you.

. . .

HELMER: But to leave your home, your husband, your children! Have you thought what people will say? . . . But this is monstrous! Can you neglect your most sacred duties? . . . First and foremost you are a wife and mother.

NORA: I don't believe that any longer. I believe that I am first and foremost a human being, like you—or anyway, that I must try to become one.

. . .

HELMER: Nora, I would gladly work for you night and day, and endure sorrow and hardship for your sake. But no man can be expected to sacrifice his honor, even for the person he loves.

NORA: Millions of women have done it.

HELMER: Oh, you think and talk like a stupid child.

NORA: That may be. But you neither think nor talk like the man I could share my life with.

. . .

NORA: I can't spend the night in a strange man's house.

HELMER: But can't we live here as brother and sister, then—?

NORA: You know quite well it wouldn't last.

. . .

HELMER: Nora—can I never be anything but a stranger to you?

NORA: Oh, Torvald! Then the miracle of miracles would have to happen.

HELMER: The miracle of miracles?

NORA: You and I would both have to change so much that—oh, Torvald, I don't believe in miracles any longer.

HELMER: But I want to believe in them. Tell me. We should have to change so much that—?

NORA: That life between us two could become a marriage. Goodbye.

The intimacy of the connection between these excerpts and the themes of the films of remarriage will not, I think, make itself felt unforgettably until one is well into the studies of the individual films; certainly, as I indicated, I did not see the intimacy until I was just about through composing them. *A Doll House* is a structure in which an apparently orderly life shatters into fragments which assemble with raging velocity an argument concerning the concepts of forgiveness, inhabitation, conversation, happiness, playtime, becoming human, fathers and husbands, brother and sister, education, scandal, fitness for teaching, honor, becoming strangers, the miracle of change, and the metaphysics of marriage. The argument of a comedy of remarriage requires, with others, each of these concepts. In *A Doll House* a woman climactically discovers that her eminently legal marriage is not comprehensible as a marriage, and therefore, before her own conscience, that she is dishonored. She demands an education and leaves to seek one that she says her husband is not the man to provide. They could find a life together (and so perhaps find or create marriage between them) only on the condition that a miracle of change take place. I have described the genre of remarriage in effect as undertaking to show how the miracle of change may be brought about and hence life together between a pair seeking divorce become a marriage. *A Doll House* thus establishes a problematic to which the genre of remarriage constitutes a particular direction of response, for which it establishes the conditions or costs of a solution.

How is this possible? Are these films as good as Ibsen's plays? But if what I have said is true about the intimacy and the exactness of the films' responses to the problematic of *A Doll House* is it important to ask whether they are as good? What is the doubt about them?

In a speech Ibsen gave to the Norwegian League for Women's Rights in 1898, he began by disclaiming the honor of having consciously worked for the women's rights movement. This disclaimer seems to encode two further claims of his opening paragraph: that the movement for women's rights is a part of the task of human advancement, whether the leading part in a given historical moment it is perhaps less important to say; and the task of human advancement he does claim to have worked for—if I understand—by saying: "I have been more the poet and less the social philosopher than people generally seem inclined to believe."* The chain of concepts I extracted from the closing pages of *A Doll House* is hardly one that an observer of society would hit upon either to describe Victorian marriage or to make a case against. it. An advocate of such marriage would have had a defense against Nora's case against it or he would have refused, unlike Torvald Helmer, to grant that a case had been made against marriage, perhaps by repeating differences between men and women which nobody need deny, and surely by saying that Nora's language—about dolls and honor and ignorance—is exaggerated, romantic. Helmer in fact takes this line in his initial responses to Nora's onslaught but soon he gives way before it, trying to comprehend her. His weakness is then humanly to his credit, his only hope for a future with her. The power of the drama lies in feeling the forming of Nora's moral conscience, her acceptance of her unprotected identity (in such lines as, "I realized that for eight years I had been living here with a complete stranger, and had borne him three children! . . . I could tear myself to pieces!"), and recognizing the con-

---

* Evert Sprinchorn, *Ibsen: Letters and Speeches* (London, MacGibbon and Kee, 1965), p. 337.

cepts of her newly created and creating consciousness, accordingly, as unanswerable.

There is in these closing pages of the play an unfolding of actions amounting to what I should like to call continual poetic justice. The intellectual or spiritual succession of concepts, dismantling the doll house, have this quality as certainly as the more obviously Ibsenist gesture in which Nora refers to her changing her clothes as her being changed, or the final sound of the slamming door of the house, which counts not as the interruption of an argument but as its continuation by other means, and specifically its ending. Her action is not the preventing but the abandoning of words, and of the house of words. The actions and words of our films characteristically work with these poetic densities—the subverted embrace at the close of *Bringing Up Baby*; the darkening screen, empty of figures, at the close of *It Happened One Night*; the photographs that close *The Philadelphia Story*; sitting down together in *His Girl Friday*; a song and dance in *The Awful Truth* . . . The Ibsen, and these films, declare that our lives are poems, their actions and words the content of a dream, working on webs of significance we cannot or will not survey but merely spin further. In everyday life the poems often seem composed by demons who curse us, wish us ill; in art by an angel who wishes us well, and blesses us.

Claiming Ibsen as well as Shakespeare as part of the specific inheritance earned by these American films, I seem to be moving toward a claim that American film is an ampler inheritor of the history of drama than American theater has been. It would be no objection to this thought to point out that three of our films have their source in American plays (two most famously, *His Girl Friday* from Hecht and MacArthur's *The Front Page*, and *The Philadelphia Story* from Philip Barry's play of that name). This is certainly to be studied, as is the issue generally of the relation of theater to film. I have not tried to do so in these pages and I make that all right with myself with the following two thoughts. First, I am not writing the history of the genre in question but proposing its logic (a distinction I will come back to). Second, more important from my point of view than locating sources is to understand what a source is. My working hypothesis throughout the following discussions is that the sources of these films bear to them no more decisive or more uniform a relation than, say, the sources of Shakespeare's plays bear to his plays. Whatever an earlier play called something like *King Lear* contains, its translation into Shakespeare's medium is inherently unpredictable; and however interesting the comparison may be in certain cases, it cannot determine what is going on in the Shakespeare. A complementary relation is that between a work of Shakespeare and certain spectaculars or panoramas "based upon" that work. In that case you might call Shakespeare's text not a source but a sea, from which various items—treasures, corpses, shells, weeds, more or less at will—were lifted and heaped on the shore of big entertainment.

I assume that the relations of source and of sea are both to be found in film, perhaps in different proportions than in other stretches of the history of drama. My purpose at the moment is to emphasize that translation into the medium of film is inherently unpredictable. A film will make of a play what it will. (In the case of the translation of human beings and of material objects this is the theme of my essay "What Becomes of Things On Film?"*) It is the film *The Philadelphia Story* and its participation in the genre of remarriage that tells you what happens to Tracy Lord's brother on film, I mean why he is incorporated into the figure of her once and future husband; the stage play has nothing to say on the subject.

I am always saying that we must let the films themselves teach us how to look at them and how to think about them. The following is a quite didactic moment that concerns the nature of a "source." It occurs at the end of *His Girl Friday* as Cary Grant phones his paper to tell Duffy, the city editor, to tear out the front page because he and Hildy are coming in with the real story. As the plot of the film is, so to speak, taking its course alongside him, Grant goes into detail about what should be taken off the front page and what left in and put where. I understand this as a fairly strict allegory of Howard Hawks telling his "re-write" man what to do with *The Front Page* (the play and the earlier film made from the play). Among other things Grant tells Duffy to do is to stick Hitler in the funny pages and to "Leave the rooster story alone. That's human interest." In part the allegory is a daring self-justification of comedy, of why one must make room for it, that what is news is determined by what human beings are humanly interested in, and you cannot know this apart from consulting that experience. Maybe it is in a rooster; and maybe Hitler is not news but just a problem about which we know what must be done. Further, it will emerge early in the reading of *The Awful Truth* that *His Girl Friday* includes elaborate allusions

* *Philosophy and Literature* 2, no. 2 (Fall 1978): 249–257.

to it, as might be expected of a film that re-casts Cary Grant and Ralph Bellamy into so similiar a position with respect to one another, and to a former wife; but only late in the reading of *The Awful Truth*, after the point about the allusions is past, do I mention that it has a good rooster story in it that in *His Girl Friday* Cary Grant, or rather Howard Hawks, is surely praising Leo McCarey for having put in. But the principal point of the allegory would be to declare that the relation between *His Girl Friday* and its "source" is one of mere practicality, that Hawks feels no more obligation or piety toward the earlier work than a managing editor would feel toward the set-up of a front page that must be re-set in the light of new and startling developments. You just have to start over again, though some of the news may well remain where it is. One may take this as an allegory confined in reference to Howard Hawks's practice in this film. To me it reads as a reasonable statement about sources generally, about one way in which a source is pressed into service. An eventual work may follow a source closely or not, in one place or another. Not every way of following amounts to an adaptation. The relation, and the purpose, will have to be made out, critically, in the individual case. I take Hawks's purpose in his allegory about sources to declare at once that his work is fresher than its reputed source and of greater human interest. (Why a given writer is drawn to particular sources is a further range of question.)

HAVING LOCATED certain causes for the genre's beginning when it does, I ought perhaps to have some speculation about why it ends when it does. It would be an answer to say that it ends when the small set of women who made it possible are no longer of an age to play its leads. Yet one feels that if the genre has not exhausted its possibilities and if the culture needs them sufficiently, people will be found. And indeed it is not clear that the genre has yielded itself up completely. Three of the most successful American films, and most interesting, of the past couple of years have begun with divorce and attempted and speculated about remarriage—*Starting Over, An Unmarried Woman,* and *Kramer vs. Kramer.* I believe *An Unmarried Woman* is generally thought to be a better film than *Starting Over,* the comparison invited by the presence in both of Jill Clayburgh as the female lead. I think the reason for that opinion is a reluctance on the part of people of a certain cultivation to see how charming and perceptive a man Burt Reynolds can be, when not cast as a good old boy. The writing of *An Unmarried Woman* may be more literate than that of *Starting Over* but in the latter film the pair's saying of words to one another is shown to mean more; their conversations are meant to bring about believable change.

And then at the climactic conclusion of *Kramer vs. Kramer*, one of the most celebrated films of 1979, exactly 100 years after the opening of *A Doll House*, one for a moment, caught in the force of Meryl Streep's performance, might have the sense that one was seeing Nora return home. The film opens with her saying to her husband who is carrying on some business over the telephone that she is leaving him and their child, going out in search of an education, in search of herself. You don't know at the close of the film whether she will stay after she goes up to see her child, but the conditions are favorable: she comes back because she is ready to be with the child, and she understands that in her absence the child and its father are at home. That on this basis a further development in the genre of remarriage can take place, one that includes the presence of children, cannot be ruled out by this film. But it cannot be ruled in either, because the film constitutes no study of these matters; we have no feeling for their lives before she left, we know nothing specific about what she has learned about herself, and we have not, except for a moment of greeting, seen her with the child. We have seen enough of the father and child's life together to want it to continue, but we have seen nothing else that we want to see resume ("only a little different this time," as Cary Grant had said to Irene Dunne some forty years before).

To assess my claim that the Hollywood sound comedy of remarriage begins with *It Happened One Night*, in 1934, one will have to know more definitely what I mean by a genre and what I mean by its having a beginning. I have already said that my date may be off—an earlier film may present itself for consideration (even one from the silent era, if a critic can show that even the fact of sound should not be regarded as essential to the genre), or it might be argued that *It Happened One Night* is for some reason not a true member of the genre, so that it only begins later, say with *The Awful Truth*. But I have also said that I am not writing history. More urgent than the date is to know what any such date should be taken to mean. My thought is that a genre emerges full-blown, in a particular instance first (or set of them if they are simultaneous), and then works out its internal consequences in further instances. So that, as I would like to put it, it has no history, only a birth and a logic (or a biology). It has a, let us say, prehistory, a setting up of the conditions it requires for viability (for example, the technology and the achievement of sound movies, the existence of certain women of a certain age, a problematic of marriage established in certain segments of the history of theater); and it has a posthistory, the story of its fortunes in the rest of the world, but all this means is that later history must be told with this new creation as a generating element. But if the genre emerges full-blown, how can later members of the genre *add* anything to it?

This question is prompted by a picture of a genre as a form characterized by features, as an object by its properties; accordingly to emerge full-blown must mean to emerge possessing all its features. The answer to the question is that later members can "add" something to the genre because there is no such thing as "all its features." It will be natural in what follows, even irresistible, to speak of individual characteristics of a genre as "features" of it; but the picture of an object with its properties is a bad one. It seems to underlie certain "structuralist" writings.

An alternative idea, which I take to underlie the discussions of this book and which I hope will be found worth working out explicitly, picks up a suggestion I broached in "A Matter of Meaning It" in *Must We Mean What We Say?* and again in *The World Viewed*, that a narrative or dramatic genre might be thought of as a medium in the visual arts might be thought of, or a "form" in music. The idea is that the members of a genre share the inheritance of certain conditions, procedures and subjects and goals of composition, and that in primary art each member of such a genre represents a study of these conditions, something I think of as bearing the responsibility of the inheritance. There is, on this picture, nothing one is tempted to call *the* features of a genre which all its members have in common. First, nothing would count as a feature until an act of criticism defines it as such. (Otherwise it would always have been obvious that, for instance, the subject of remarriage was a feature, indeed a leading feature, of a genre.) Second, if a member of a genre were just an object with features then if it shared *all* its features with its companion members they would presumably be indistinguishable from one another. Third, a genre must be left open to new members, a new bearing of responsibility for its inheritance; hence, in the light of the preceding point, it follows that the new member must bring with it some new feature or features. Fourth, membership in the genre requires that if an instance (apparently) lacks a given feature, it must compensate for it, for example, by showing a further feature "instead of" the one it lacks. Fifth, the test of this compensation is that the new feature introduced by the new member will, in turn, contribute to a description of the genre as a whole. But I think one may by now feel that these requirements, thought about in terms of "features," are beginning to contradict one another.

(Before articulating that feeling I pause for an aside to readers of Wittgenstein's *Philosophical Investigations* who will sense a connection here, in the denial that what constitutes the members of a genre is their having features in common, with Wittgenstein's caution not to say of things called by the same name that they must have something in common [hence share some essence or so-called universal] but instead to consider that they bear to one another a family resemblance. But if I said of games, using Wittgenstein's famous example in this connection, that they form a genre of human activity, I would mean not merely that they look like one another or that one gets similar impressions from them; I would mean they *are what they are* in view of one another. I find that the idea of "family resemblance" does not capture this significance, if indeed it is really there.)*

Take an example. I have mentioned that one feature of the genre of remarriage will be the narrative's removal of the pair to a place of perspective in which the complications of the plot will achieve what resolution they can. But *It Happened One Night* has no such settled place; instead what happens takes place "on the road." I say that what compensates for this lack is in effect the replacement of a past together by a commitment to adventurousness, say to a future together no matter what. But then it will be found that adventurousness in turn plays a role in each of the other films of remarriage. And one may come to think that a state of perspective does not require representation by a place but may also be understood as a matter of directedness, of being on the road, on the way. In that case what is "compensating" for what? Nothing is lacking, every member incorporates any "feature" you can

* I am prompted to these parenthetical remarks by an exchange of letters with Paul Alpers.

name, in its way. It may be helpful to say that a new member gets its distinction by investigating a particular set of features in a way that makes them, or their relation, more explicit than in its companions. Then as these exercises in explicitness reflect upon one another, looping back and forth among the members, we may say that the genre is striving toward a state of absolute explicitness, of expressive saturation. At that point the genre would have nothing further to generate. This is perhaps what is sometimes called the exhaustion of conventions. There is no way to know that the state of saturation, completeness of expression, has been reached.

A NATURAL QUESTION ARISES as to how comedies of remarriage are related to films in which the fact of remarriage can be said to be dominant but the film is not a comedy. A good case is *Random Harvest* (Mervyn LeRoy, 1942), with Ronald Coleman and Greer Garson. This is complete with divorce; with spiritual death and revival; with the question of whether the man or the woman is the active member of the pair; with discussions of life as beginning with the meeting of the pair, the past having nothing in it but their past; with the return to a particular house in the country which holds the key to a saving perception—all matters that turn out to be part of the grain of remarriage comedies. But obviously this romance, despite its locating a certain happiness, is all wrong for our genre, somehow its opposite. It does not explain this fact to say that *Random Harvest* is not a comedy; it reasserts the fact. The question is how the films of remarriage add up such similar events to so dissimilar an effect. The difference cannot be expressed as a difference in the explicitness of features for which the relation of compensation can make up, since there is at least one feature absent from *Random Harvest*—the man never claims the woman, never declares his right to her desire—for which there is no compensation. It seems to me rather that this absence *negates* something necessary to the genre of remarriage.

The truth of these assertions aside for the moment (they cannot be assessed apart from the readings of the films to come), the idea of negation in contrast to that of compensation here suggests a way to express the intuition I have of how to think about films related to one another not as members of the same genre but as members of adjacent genres. Let us think of the common inheritance of the members of a genre as a story, call it a myth. The members of a genre will be interpretations of it, or to use Thoreau's word for it, revisions of it, which will also make them interpretations of one another. The myth must be constructed, or reconstructed, from the members of the genre that inherits it, and since the genre is, as far as we know, unsaturated, the construction of the myth must remain provisional. Before seeing how a construction might go, I note that a minor member of a genre may hit upon a startling interpretation or revision of a passage of the myth. The central idea of *Remember?* (Norman Z. McLeod, 1939), with Robert Taylor and Greer Garson, is to interpret the passage about renewal as a story of starting again without knowledge, a condition it depicts as produced by an amnesia-producing drug. This in effect interprets the idea of a love potion—of whatever the thing is that makes love possible, or recognizable—as providing the gift of pastlessness, allowing one to begin again, free of obligation and of the memory of compromise. But let us see how the general construction of the myth might go.

A running quarrel is forcing apart a pair who recognize themselves as having known one another forever, that is from the beginning, not just in the past but in a period before there was a past, before history. This naturally presents itself as their having shared childhood together, suggesting that they are brother and sister. They have discovered their sexuality together and find themselves required to enter this realm at roughly the same time that they are required to enter the social realm, as if the sexual and the social are to legitimize one another. This is the beginning of history, of an unending quarrel. The joining of the sexual and the social is called marriage. Something evidently internal to the task of marriage causes trouble in paradise—as if marriage, which was to be a ratification, is itself in need of ratification. So marriage has its disappointment—call this its impotence to domesticate sexuality without discouraging it, or its stupidity in the face of the riddle of intimacy, which repels where it attracts, or in the face of the puzzle of ecstasy, which is violent while it is tender, as if the leopard should lie down with the lamb. And the disappointment seeks revenge, a revenge, as it were, for having made one discover one's incompleteness, one's transience, one's homelessness. Upon separation the woman tries a regressive tack,

usually that of accepting as a husband a simpler, or mere, father-sub-
stitute, even one who brings along his own mother. This is psychologically an effort to put her desire, awakened by the original man, back to
sleep . . .

We would have to continue the story by telling the role of the pair's
fathers and mothers and of the possibility of their having children. Let
us not anticipate what the films themselves will have to say about these
matters. And let us assume that the quarrel is going to have to take up
questions about who is active and who is passive, and about who is
awake, and about what happiness is and whether one can change. The
quarrel, the conversation of love, takes lavish expenditures of time, exclusive, jealous time; and since time is money, it requires a way to understand where the (man's) money comes from to support so luxurious
a leisure. The pair is attractive, their wishes are human, their happiness
would make us happy. So it seems that a criterion is being proposed for
the success or happiness of a society, namely that it is happy to the extent that it provides conditions that permit conversations of this character, or a moral equivalent of them, between its citizens. Then the ending
clarifies these themes by deepening the mystery of the pair's connection. It is the man's turn to make the move—the woman had presumably started things with something called an apple, anyway by presenting a temptation. The man must counter by showing that he has
survived his yielding and by finding a way to enter a claim. To make a
correct claim, to pass the test of his legitimacy, he must show that he is
not attempting to command but that he is able to wish, and consequently to make a fool of himself. This enables the woman to awaken
to her desire again, giving herself rather than the apple, and enables the
man to recognize and accept this gift. This changing is the forgoing or
forgetting of that past state and its impasse of vengefulness, a forgoing
symbolized by the initial loss of virginity.

In the construction of the myth, the picture of the properties of an
object is replaced by an idea of the clauses or provisions of a story.
Then to say that, to recur to my former instances, adventurousness
compensates for the provision concerning a location of perspective is to
say that the concept of adventurousness is an interpretation of the same
story, allows it to go on being told, being developed; the genre remains
the same, it is further defined. Whereas to say that the man's inability

to claim the woman negates that provision is to say that it changes the story; the genre is different, an adjacent genre is defined. Which of these is true of a given film and its interpretations cannot be decided at a glance. The consequences have to be followed out. In *Random Harvest*, the absence of the claim goes with events that require not merely the absence but the denial of the possibility of children for the marriage, and it means (consequently?) the withholding of sexual gratification during a dozen or so years of what is called the prime of life—anyway until after the age of child-bearing. (Quite as if we have here a participant in a genre whose myth presents a punishment for living the myth of remarriage, or for failing it.) Both compensation and negation, as procedures of what Thoreau calls "revising Mythology," are terms in which he might have described his life of writing as such. Another way to characterize what I called earlier the exhaustion of convention or the saturation of expressiveness is to say that when a myth can no longer support revision—the being looked over again—then the myth has died, we have died to it. (If the notion of dialectic meant much to us we might note the dialectical leanings of words like compensation and negation. A clause is neither just satisfied nor just unsatisfied but is satisfied or unsatisfied in some way, in some aspect, say literally or abstractly or ironically or individually . . . This [partial] satisfaction then changes the issue, which then must press on for further satisfaction, if the issue is still living.)

The concept of adjacent genres is something for future work. The principal other explicit call upon it in the book occurs late in Chapter 6, on *Adam's Rib*, at the end of an excursus on some related films of George Cukor. An implied contrast is thus set up between the concept of a genre and the concept of an oeuvre; the ground of the contrast seems to be that the latter, unlike the former, is meant to account for an historical order among its members. This contrast between genre and oeuvre prompts me to mention an essay I have just completed on Hitchcock's *North by Northwest** which locates this film at the same time within the development of Hitchcock's oeuvre and adjacent to the structure of the genre of remarriage. Specifically, the fact that it is the man, and not the woman, who undergoes something like death and re-

* Forthcoming in *Critical Inquiry*.

vival seems to be what allows the pair (uniquely in Hitchcock's romantic adventures) to be shown to marry, and in negating a clause of the myth of the genre of remarriage, the film declares its own way of working out the legitimizing of marriage.

IT WILL BE EXPECTED, from what I have been saying, that the order in which I take up the reading of the major films of the genre of remarriage is meant neither as historical (in whatever sense a genre may be said to have a history) nor dialectical (since that would entail deriving the genre along with all the genres of film, a task which is hardly yet a dream). The order has rather been determined by the practical or strategic problems of exposition. Having found that *The Lady Eve* made for a reasonably clear sketch both of the generic and the Shakespearean dimensions of the task I set myself, I wanted to follow it, as a kind of second beginning, with a reading that would go as far in invoking consecutive philosophical exposition as the present book requires and permits itself to go. Hence the essay on *It Happened One Night*. The material on *Bringing Up Baby*, the first of the essays written, was called for next by a remark in the essay on *It Happened One Night*. The order of the remaining four essays was negotiated amicably. I felt the need to reaffirm immediately, in as it were a third beginning, the theme of remarriage, after such fierce displacements of it. In *Adam's Rib* it is also displaced, so the fourth place in the readings would have to go to one of the other three films. For reasons that I hope make themselves plain in the essay on *The Awful Truth*, I felt that film should come last. *His Girl Friday* I wanted to follow *The Philadelphia Story*, with which it makes a pair; hence *The Philadelphia Story* comes fourth, putting *Adam's Rib* sixth. These last four essays, in contrast to the first three, were written knowing that the others were, or would be, written, and knowing what they looked like; that is, knowing that I was writing a book.

These facts are consequential. Once the fourth essay was done it locked the preceding ones in place, more or less in their original shapes, and became the site from which the essay to follow could survey its visible tasks, itself in turn becoming a site . . . So while the genre may not care, so to speak, in what order its instances are generated, a book about the genre is affected at every turn by the order it imposes upon itself. The essays are quite different from one another and it is clear to me that each of the readings would bear a different countenance had its order in the composition of the essays been different. Does this impugn the objectivity of my readings?

Two worries, trenched on in the course of this introduction, are generally expressed concerning critical objectivity: that a critic is reading something into a text; and that there may be more than one interpretation of a text. I mention them because nowadays it is equally in fashion, in other circles, to say that objectivity is neither possible nor desirable (like being a mermaid), and that far from seeking *one* interpretation of a text we should produce as many as our talents will allow. The watchword should be fun. In making a couple of concluding remarks about these worries I emphasize that the most important fact about them, to my mind, is their unclarity; so that the best instruction the worries have for us lies in trying to describe that very unclarity.

The idea of reading something into a text seems to convey a picture of putting something into a text that is not *there*. Then you have to say what *is* there and it turns out to be nothing but a text. But in *that* sense you might just as well say that there is no dog in the text "Beware of the dog." Is *that* what someone feels the error of overreading to be, a relatively simple, psychotic notion that an animal, for example, is a word? Naturally I do not deny that some readings are irresponsible in fairly straightforward ways. But "reading in," as a term of criticism, suggests something quite particular, like going too far even if on a real track. Then the question would be, as the question often is about philosophy, how to bring reading to an end. And this should be seen as a problem internal to criticism, not a criticism of it from outside. In my experience people worried about reading in, or overinterpretation, or going too far, are, or were, typically afraid of getting started, of reading as such, as if afraid that texts—like people, like times and places—mean things and moreover mean more than you know. This is accordingly a fear of something real, and it may be a healthy fear, that is, a fear of something fearful. It strikes me as a more discerning reaction to texts than the cheerier opinion that the chase of meaning is just as much fun as man's favorite sport (also presumably a thing with no fear attached). Still, my experience is that most texts, like most lives, are underread, not overread. And the moral I urge is that this assessment be made the subject of arguments about particular texts.

As for the claim that there are interpretations other than the ones I give, let me be quick not just to avoid the impression of denying this, as though I were eager to be known as a tolerant liberal on this issue; let me prove that there *must* be more than one interpretation possible. Call the reading I give of a film the provision of a text about a text. Think of this provision as a secondary text and let us say that it is an interpretation of the primary one. Then, among other things, we owe an account of what an interpretation is. I pick up the suggestion from Wittgenstein's celebrated study, in Part II of *Philosophical Investigations*, that what he calls "seeing an aspect" is the form of interpretation: it is seeing something *as* something. Two conditions hold of a case in which the concept of "seeing as" is correctly employed. There must be a competing way of seeing the phenomenon in question, something else to see it as (in Wittgenstein's most famous case, that of the Gestalt figure of the "duck-rabbit," it may be seen as a duck or as a rabbit); and a given person may not be able to see it both ways, in which case it will not be true for him that he sees it (that is, sees a duck or sees a rabbit) *as* anything (though it will be true to say of him, if said by us who see both possibilities, that he sees it as one or the other). And one aspect dawns not just as *a* way of seeing but as a way of seeing something now, a way that eclipses some other, definite way in which one can oneself see the "same" thing.

Accordingly, taking what I call readings to be interpretations, I will say: for something to be correctly regarded as an interpretation two conditions must hold. First there must be conceived to be competing interpretations possible, where "must" is a term not of etiquette but of (what Wittgenstein calls) grammar, something like logic. Hence to respond to an interpretation by saying that there must be others is correct enough but quite empty until a competing interpretation is suggested. Second, a given person may not be able to see that an alternative is so much as possible, in which case he or she will not know what it means to affirm or deny that an interpretation involves reading in, hence will have no concrete idea whether one has gone too far or indeed whether one has begun at all. So many remarks one has endured about the kind and number of feet in a line of verse, or about a superb modulation, or about a beautiful diagonal in a painting, or about a wonderful camera angle, have not been readings of a passage at all, but something like items in a tabulation, with no suggestion about what is being counted or what the total might mean. Such remarks, I feel, *say* nothing, though

they may be, as Wittgenstein says about naming, preparations for saying something (and hence had better be accurate). The proof that there
must be competing interpretations speaks to two recurrent issues. It
helps one see why someone wishes to insist, more or less emptily, that
there *must* be another interpretation: since one interpretation eclipses
another it may present itself as *denying* the possibility of that other. It
also helps me see what a complete interpretation could be, how it is one
may end a reading. Completeness is not a matter of providing *all* interpretations but a matter of seeing one of them *through*. Reading in,
therefore, going too far, is a risk inherent in the business of reading,
and venial in comparison with not going far enough, not reaching the
end; indeed it may be essential to knowing what the end is.

Having now spoken of my readings as secondary texts and described
them as interpretations, I would like to propose an alternative to the
concept of interpretation as a mode of describing these texts—which is
to say: I would like to start providing a tertiary text. There are many
such tertiary passages in the discussions to follow and, having said that
such a notion of a hierarchy of texts creates obligations of explanation,
let me at least note that it is not clear that these levels mean the same
thing. A tertiary text, as I just introduced the term, is just a text referring to itself, and not all ways of referring to itself are departures from
itself. So maybe there is no higher text (of reading) than a secondary
one. But secondary, then, as opposed to what? Is the primary thing a
text in the same sense? Suppose that an interpretation just is of a text
and that to be a text just is to be subject to interpretation; and suppose
this means that a text constitutes interpretation. A secondary text is a
text in admitting of an interpretation but also in being an interpretation
of a text. Is a primary text an interpretation of a text? Unless we see how
what it interprets is a text (for example, how the world, or a person, is a
text) we may not know how it is a text.

This aside, what I was going to call my tertiary text, my alternative to
speaking of interpretation, is this: A performance of a piece of music is
an interpretation of it, the manifestation of one way of hearing it, and it
arises (if it is serious) from a process of analysis. (This will no longer be
the case where a piece just *is* its performance; where, say, it is itself a
process of improvisation.) Say that my readings, my secondary texts,
arise from processes of analysis. Then I would like to say that what I am
doing in reading a film is performing it (if you wish, performing it in-

side myself). (I welcome here the sense in the idea of performance that it is the meeting of a responsibility.) This leaves open to investigation what the relations are between performance and interpretation, and between both of these and analysis, and between differing analyses, and hence between differing performances.

THE WAY I HAVE SPOKEN of interpretation (marked by the occurrence of a certain use of "as," that is, of comparison, of a point of view dawning) is meant to mark a significant relation between the thought of *Philosophical Investigations* and the thought of Heidegger's *Being and Time*. Further relations between these writings are pointed to in the remarks entitled "Film in the University" that I have placed in the Appendix. These remarks were written as the introductory half of an essay the second half of which consisted of the reading of *Bringing Up Baby* that appears as Chapter 3. I retain those introductory remarks here if for no other reason than that they say things not said elsewhere in this book about who I am, I mean who I is, who the I in this book is, how that figure thinks things over and why such a one takes film as something to think over.

There is another reason for retaining them. That introduction was written as part of the opening address of a conference entitled Film and the University.* The initiating and recurrent topics of the conference had to do with what was (and is) called the legitimacy of film study. However one conceives of this issue, I am for myself convinced that a healthy future of film culture, hence of useful, orderly, original film criticism and theory, is as bound to film's inhabitation of universities (whatever universities in turn have come to be, and will further come to be because of that inhabitation) as was the epochal outburst of American literary criticism and theory that produced the New Criticism of an earlier generation. But my hope for the future of film culture is not based on that healthy development alone, and the ambition of this book is not limited to wishing a role for itself in that development.

The hope and the wish are based as well on the fact that films persist as natural topics of conversation; they remain events, as few books or

---

* Organized by Marshall Cohen and Gerald Mast and held in July 1975 at the Graduate Center of the City University of New York.

plays now do. I would like that conversation to be as good as its topics deserve, as precise and resourceful as the participants are capable of. I would like, to begin with, conversations about movies, and therefore daily or weekly reviews of them, to be as uniformly good as we expect conversations or columns about sports to be. Not as widespread, perhaps, if that matters. My fantasy here is of conversations about *It Happened One Night*—or, for that matter, about *Kramer vs. Kramer*—that demand the sort of attention and the sort of command of relevant facts that we expect of one another in evaluating a team's season of play; conversations into which, my fantasy continues, a remark of mine will enter and be pressed and disputed until some agreement over its truth or falsity, some assessment of its depth or superficiality, has been reached.

This is a fantasy any writer may at any time harbor about being read attentively; but it is also a fantasy that could only recently have become practical about movies. It depends on a certain access to at least some parts of the history of film, a fateful development I described earlier as increasingly at hand. But if the conversation, the culture I fantasize, is technically at hand, something further, something inner, untechnical, keeps it from our grasp.

We seem fated to distort the good films closest to us, exemplified by the seven concentrated on in this book. Their loud-mouthed inflation by the circus advertising of Hollywood is nicely matched by their thin-lipped deflation by those who cannot imagine that products of the Hollywood studio system could in principle rival the exports of revolutionary Russia, of Germany, and of France. This view sometimes seems the work of certain critics or scholars of film with a particular anti-American axe of contempt to grind. But it expresses, it feeds on, a pervasive conflict suffered by Americans about their own artistic accomplishments, a conflict I have described elsewhere as America's overpraising and undervaluing of those of its accomplishments that it does not ignore.* It is part of this situation that American film directors play to it. The case of Howard Hawks comes to mind. The films of his discussed in this book seem to me clearly the work of a brilliant, educated, if brutal, mind, and one that knows its craft; the work, you might well say, of an artist. Yet in the interviews Hawks submitted to upon his dis-

---

* *The Senses of Walden* (New York: The Viking Press, 1972), p. 33.

covery by educated circles a decade or so ago, he presents himself as a cowboy. I assume this is a natural extension of his brilliance and education and brutality. It is as if he knew that for an American artist to get and to keep hold of a public he must not be perceived as an artist, except on condition; above all he must not seem to recognize himself as such. The condition that would take the curse off his claiming to be an artist is that he seem so weird that no person of reasonably normal tastes could be expected to want to pay the price of being like him.

It is complicated. Part of Orson Welles's reputed troubles with Hollywood was that he carried the air of an artist, or a genius, or something like that. But as if in compensation his clientele apparently accepts his work—*Citizen Kane* at least—as a work of genius and of art. I find it a dangerous model for naming such aspirations. It seems to me that what is being called art in that work is showmanship and that what is good in the film may not depend on its overt showmanship. It would follow that the craft lies in its effects, not in its basis; that the workmanship is arbitrary, not authoritative. This is not to deny that great artists may sometimes be great showmen, nor even to deny that something you might call showmanship is essential to major art, as active in Emily Dickinson as in Walt Whitman. While we're at it, take two showmen like Eisenstein and Frank Capra. The former is an intellectual, the latter is not, but as craftsmen they seem to me to resemble one another, especially in putting things together for their melodramatic value. Either might have hit, for example, on Edward Arnold and his cigars and diamond rings as the image of a capitalist munitions maker. (Both knew some Dickens.) This conjunction of minds will seem preposterous to some who care about film, to some partisans of each of them. A good reason for this feeling is the idea that Capra is not remotely as interesting visually as Eisenstein, along with an idea that film is a visual medium. Certainly it is true that nothing in Capra could satisfy an interest in the visual, in what one might call the melodramatically visual, the way Eisenstein can by, for example, watching the carcass of a horse drop from an opening drawbridge into the water far below. But suppose film's interest in the visual can be understood as a fascination with the fact of the visible. Then nothing in Eisenstein could be more revealing than Capra's camera, in *It's a Wonderful Life,* in the sequence in which James Stewart, greeting his returning brother at the railroad station, learns that this return does not mean his release from his hated obligations but his final sealing within them, as it accompanies Stewart's circling away from the scene of happy exchanges, reeling from the collapse of his ecstasy, working to recover himself sufficiently to find a public face. We are vouchsafed a vision of the aging American boy, as melodramatically private as a Czar.

Philistines about film may take reassurance from such observations about Hawks and Welles and about the comparison of Eisenstein and Capra. That would be because they are philistines, who prefer reassurance to all things. A significant worry for me is that sophisticates about film may regard the same remarks as heresies. As heresies the observations are uninteresting, which means to me that the orthodoxies are equally uninteresting which cast them as heresies. My worry is that instead of such issues becoming examples of the ongoing conversation about film I was fantasizing (which is what they are designed to be), the orthodoxies will receive tenure in university programs of study, and therewith unnatural leases on life. What then? Should one try to convince oneself that universities are not as urgent for the future of film studies as I have taken them to be? Not to strike even though the iron is hot is sometimes the creative way to proceed. But it is of limited value as a general principle of conduct. (I distinguish this from the more popular principle of striking while the hammer is hot.)

But there is something beyond our distorting of the value of the good films closest to us that keeps them inaccessible to us as food for thought. It lies in the dilemmas I was invoking in calling upon Emerson's appeal to the common and the low, and his and Thoreau's passion for the near, claiming their affinity with my philosophical preoccupation with the ordinary, the everyday. The dilemmas concern what I called taking an interest in one's experience. The films that form the topics of the following chapters are ones some people treasure and others despise, ones which many on both sides or on no side bear in their experience as memorable public events, segments of the experiences, the memories, of a common life. So that the difficulty of assessing them is the same as the difficulty of assessing everyday experience, the difficulty of expressing oneself satisfactorily, of making oneself find the words for what one is specifically interested to say, which comes to the difficulty, as I put it, of finding the right to be thus interested. It is as if we and the world had a joint stake in keeping our-selves stupid, that is dumb, inarticulate. This poses, to my mind, the specific difficulty of philosophy and calls upon its peculiar strength, to receive inspiration for taking thought from the very conditions that op-pose thought, as if the will to thought were as imperative as the will to health and to freedom.

. . . perhaps the finest, strongest, happiest, *most courageous* period of Wagner's life: the period during which he was deeply concerned with the idea of Luther's wedding. Who knows upon what chance events it depended that instead of this wedding music we possess today *Die Meistersinger?* And how much of the former perhaps still echoes in the latter? But there can be no doubt that "Luther's Wedding" would also have involved a praise of chastity. And also a praise of sensuality, to be sure—and this would have seemed to be quite in order, quite "Wagnerian."

For there is no necessary antithesis between chastity and sensuality; every good marriage, every genuine love affair, transcends this antithesis. Wagner would have done well, I think, to have brought this *pleasant* fact home once more to his Germans by means of a bold and beautiful Luther comedy, for there have always been and still are many slanderers of sensuality among the Germans; and perhaps Luther performed no greater service than to have had the courage of his *sensuality* (in those days it was called, delicately enough, "evangelical freedom").

NIETZSCHE, Genealogy of Morals

Had I had faith I should have remained with Regine.

KIERKEGAARD, Journals

The finding of an object is in fact the refinding of it.

FREUD, Three Essays on the Theory of Sexuality

Hegel remarks somewhere that all facts and personages of great importance in world history occur, as it were, twice. He forgot to add: the first time as tragedy, the second as farce.

MARX, The Eighteenth Brumaire

The life of man is the true romance, which, when it is valiantly conducted, will yield the imagination a higher joy than any fiction.

I have seen an individual, whose manners, though wholly within the conventions of elegant society, were never learned from there, but were original and commanding, and held out protection and prosperity; one who did not need the aid of a court-suit, but carried the holiday in his eye; who exhilarated the fancy by flinging wide the doors of new modes of existence; who shook off the captivity of etiquette, with happy, spirited bearing, good-natured and free as Robin Hood; yet with the port of an emperor,—if need be, calm, serious, and fit to stand the gaze of millions.

Since our office is with moments, let us husband them.

Our moods do not believe in each other.

EMERSON, "New England Reformers," "Manners," "Experience," "Circles"

I should like to say that what dawns here lasts only as long as I am occupied with the object in a particular way.

WITTGENSTEIN, Philosophical Investigations

1

CONS AND PROS

The Lady Eve

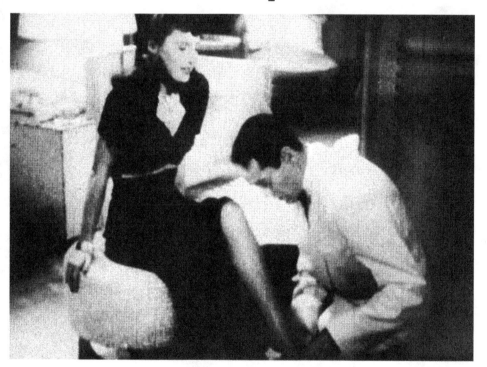

It is a leading thought of mine about the film comedies of remarriage that they each have a way of acknowledging the issue of the relation of actor and character, in particular that each has a way of harping on the identity of the real woman cast in its principal pair.

WE can make a start in reading *The Lady Eve* (1941) without considering its generic allegiances and their Shakespearean background. From the name in its title and from the animated title cards we know that Preston Sturges is going to present us with some comic version of the story of the expulsion from the Garden of Eden. And sure enough, the film opens with a young man and his guardian shadow leaving a tropical island on which he has been devoted to what he calls the pursuit of knowledge. That something is mixed up in this knowledge is confirmed at once by the camera's drifting, as if bored, or embarrassed, away from his delivery of his farewell speech declaring the purity of his pursuit, to discover his shadow leading a nubile native down to the shore, the pair sporting chains of flowers. This line of story is picked up as the leading lady attracts the young man's attention by clunking him on the head with an apple, and again by the intimate enmity revealed between the man's snake and the woman's dreams. Their relationship is broken, anyway their plans are, by the man's coming into new knowledge. As if this were not enough, we are shown the fall of the man repeated over and over, and the idea of "falling" is explicitly and differently interpreted by each of two characters (Curly and Eve), as if daring us to interpret it for ourselves. This line of argument has a most satisfactory conclusion in the man's closing declaration that he does not want to know. Had our common ancestor said that in the beginning, there would be no question of endings.

But even if we consider that Shakespearean romance itself tends to invoke the myth of Eden, such considerations merely scratch at the surface iconography of this film. Of course these considerations also merely pick up superficial items in the myth of Eden, or pick them up superficially. The myth is after all about the creation of woman and about the temptability of man. Now *The Lady Eve* is about a con artist (Barbara Stanwyck, introduced as Jean, reintroduced as Eve) who calls herself, because she is a woman, an adventuress; and it is equally about the gullibility of a man (Henry Fonda), forever being called a mug or a sucker. (Jean's name for him is Hopsy; Eve's name for him, and the world's, is Charles.) But can the film be seen to be about the creation of a woman?

Jean's central con of this man requires her reappearance as the Lady Eve. Her associate Curly (Eric Blore)—"Sir Alfred McGlennon Keith at the moment"—explains her (re)appearance by telling Hopsy/Charles the story of Cecilia, or The Coachman's Daughter, filling him, as Sir Alfred puts it, with "handsome coachmen, elderly earls, young wives, and two little girls who look exactly alike"—that is, with the very farcing of romance. Jean as Eve continues Sir Alfred's image by asking, "You mean he actually swallowed that?" and is told, "Like a wolf"—as though this story was the biggest of the fruits of the tree of knowledge that he was to be handed. Eve has·her own explanation of Charles's readiness to accept her story (that is, to accept her as not Jean), namely, that they really do not look the same to one another as they did on the boat because they no longer feel as they did then about one another. On the boat, she says, "we had this awful yen for one another." Whatever the psychological or philosophical validity and interest of her explanation, it is a fragment of a reasonable view of what romance is. Quoting one editor of *The Tempest*: "For romance deals in marvelous events and solves its problems through metamorphoses and recognition scenes—through, in other words, transformations of perception."* By the time the pair find themselves alone again, riding horses through wooded paths, compelled by the beauty of a sunset to dismount and look, and the man has begun repeating his own self-declared romance to the woman (a line of story he had originally feigned to criticize as "dull as a drugstore novel")—a repetition even the horse tries to tell him is inappropriate—by that time it may dawn on us that Preston Sturges is trying to tell us that tales of romance are inherently feats of cony catching, of conning, making gulls or suckers of their audience, and that film, with its typical stories of love set on luxurious ships or in

* Robert Langbaum, introduction to the Signet edition.

mansions and containing beautiful people and horses and sunsets and miraculously happy endings, is inherently romantic.

Granted, then, that this film does invite us to consider the source of romance, what is the implication? That we, as the audience of film, are fated, or anyway meant, to be gulled by film, including this film? This makes our position seem the same as Hopsy/Charles's. But don't we also feel that our position is equally to be allied with the woman's, at the man's expense? Who are these people and what are their positions?

Let us approach them by getting deeper into this film's identifying of itself with the tradition of romance. Take first the feature of the action's moving from a starting place of impasse to a place Frye calls "the green world," a place in which perspective and renewal are to be achieved. In *A Midsummer Night's Dream* this place is a forest inhabited by fairies, explicitly a place of dreams and magic; in *The Winter's Tale* the place is the rural society of Bohemia; in *The Merchant of Venice* the equivalent of such a place contains oracular caskets; in *The Tempest* it occupies the entire setting of the action, with the framing larger world supplied by narrative speeches; in *Bringing Up Baby, The Awful Truth,* and *Adam's Rib,* in addition to *The Lady Eve*—that is, in more than half of the definitive remarriage comedies of Hollywood—this locale is called Connecticut. Strictly speaking, in *The Lady Eve* the place is called "Conneckticut," and it is all but explicitly cited as a mythical location, since nobody is quite sure how you get there, or anyway how a lady gets there. This is Preston Sturges showing off at once his powers of parody and his knowledge of his subject, and giving us fair warning: in his green world the mind or plot will not only not be cleared and restored, it will be darkened and frozen.

Another feature of Shakespeare's late romances is an expansion of the father-daughter relationship. (This goes together with the fact that these late plays emphasize the reconciliation of an older generation at the expense of a central interest in the plight of young lovers. The comedy of remarriage is a natural inheritor of this shift of interest away from the very young.) *The Lady Eve* emphasizes the father-daughter relationship as strongly as *It Happened One Night* and *The Philadelphia Story* do, but it goes quite beyond its companions in the genre by endowing its father—as Shakespeare endows a number of his late fathers—with the power, or to use Shakespeare's word, with the art, of magic. That Harry's use of cards is meant to stand for a power possessed not merely by a shark but by a magus is declared as he sits on his daughter's bed dressed in a wizard's robe, deals "fifths" for her admiration, describing the trick as "just virtuosity," saying "you don't really need it"; and as she thereupon asks him, in their tenderest moment together, to tell her her fortune, as if for her to ask this man for a professional reading of the cards is to ask him for his blessing. In *All's Well That End's Well* the heroine has inherited her father's book and knack of magic, which proves to be the key to the happiness she is awarded. The most famous of Shakespeare's father-magicians is the central figure of *The Tempest*, the play in which renewal or reconciliation or restoration is shown to exact the task of the laying aside of magic. I understand an allusion to this task of Prospero's, or a summary of it, when Jean returns to her cabin after a day with Hopsy and announces their love for one another to her father, declaring that she would give anything to be—that she is going, she corrects herself, to be—everything he thinks she is, everything he wants her to be (a declaration coming the day after she had created and destroyed for him the wisdom of having an ideal in an object of love), and then saying to her father, "And you'll go straight too, Harry, won't you?" "Straight to," he asks, "where?" (This mode of allusion or summary might be compared with another allusion from our genre to *The Tempest* that amounts, to my ear, almost to an echo. I am thinking of the late moment of awakening in *The Philadelphia Story*—comparable to the late moment at which Miranda more fully realizes the imminence of her departure into human womanhood and human relationship, exclaiming "How beauteous mankind is!"—at which Katharine Hepburn says, in a sudden access of admiration, "I think men are wonderful.")

I do not require immediate acceptance of Hollywood fast talk as our potential equivalent of Shakespearean thought, and yet I will have at some stage to ask attention for at least one further moment of thematic coincidence between *The Lady Eve* and *The Tempest*, the coincidence of their conclusions in an achievement of forgiveness. Such attention would mean nothing to my purpose apart from a live experience of the film within which it holds its own against the Shakespearean pressure. I mean, at a minimum, that we are to ponder the experience of this man's and this woman's concluding requests to one another to be forgiven; that this bears pondering. Two ways not to bear it are either to conclude that their treatment of one another has been unforgiveable, in which case the ending of the film is either cynical or deluded by the ideology of Hollywood; or to conclude that there is no outstanding problem since human beings are fated so far as they have progressed, politically or privately, to cynicism, insincerity, and delusion in their relations with one another, above all in their dealings with love and marriage, so the film is after all realistic in its assessment both of their needing forgiveness and of their incapacities to grant it.

The unacceptability, or instability, of each of these conclusions (each gnawed at by the other) is a reason, I believe, that a typical reaction to such films is to develop a headache. (Then it may be such a reaction that produces the title "madcap comedies" for such films.) We are not yet ready to try to think our way beyond this reaction, but I mention that Frye calls particular attention to the special nature of the forgiving and forgetting asked for at the conclusion of romantic comedy: "Normally, we can forget in this way only when we wake up from a dream, when we pass from one world into another, and we often have to think of the main action of a comedy as 'the mistakes of a night,' as taking place in a dream or nightmare world that the final scene suddenly removes us from and thereby makes illusory."* *Bringing Up Baby* and *Adam's Rib* also explicitly climax or conclude with a request for forgiveness, and *The Awful Truth* and *The Philadelphia Story* do so implicitly. *His Girl Friday* notably does not, which is a way of understanding the terrible darkness of that comedy.

I should perhaps pause, still barely inside the film, to say that I am not claiming that these films of remarriage are as good as Shakespearean romantic comedies. Not that this is much of a disclaimer: practically nothing else is as good either. But I am claiming a specificness of inheritance which is itself more than enough of a problem to justify. Another two sentences from Frye will locate my claim: "All the important writers of English comedy since Jonson have cultivated the comedy of manners with its realistic illusion and not Shakespeare's romantic and stylized kind ... The only place where the tradition of Shakespearean romantic comedy has survived with any theatrical success is, as we should expect, in opera.** I am in effect adding that the Shakespearean tradition also survives in film (thus implying that film

---

* *A Natural Perspective* (New York: Columbia University Press, 1965), p. 128.
** Ibid., pp. 24, 25.

may provide an access for us to that tradition) and adding as well that such claims are all but completely up in the air, and will pretty much go on being left in the air by what I have been or will be saying here. The claim sits in the quality of one's experience of film, in the nature of film, and I am at best assuming that experience and that nature, and preparing the ground for inviting the experience of others.

I have sometimes found it useful to think of the nature of film by comparing what camera and projection bring to a script with what music brings to a libretto. Whatever the strains of the comparison its point here would be to locate what it is, in any medium that can seriously be thought of as maintaining a connection with Shakespeare's plays, that bears the brunt of his poetry. The signal weakness in comparing the poetry of the camera (of, so to speak, the photogenetic poetry of film itself) with the music of opera is that this misses, as the comparison of film with theater generally misses, the mode of uniqueness of the events on the screen. Plays may be variously produced, and productions may or may not change in the course of a run, and may be revived; films can only be rerun or remade. You can think of the events on a screen equally as permanent and as evanescent. The poetry of the final appeals for forgiveness in *The Lady Eve* is accordingly a function of the way just this man and this woman half walk, half run down a path of gangways, catching themselves in an embrace on each landing, and how just this sequence of framings and attractions of the camera follow these bodies as they inflect themselves to a halt before a closed door, and just the way these voices mingle their breaths together. These moments are no more repeatable than a lifetime is. The uniqueness of the events of film is perhaps better thought of in comparison with jazz than with opera. Here the point of contact is that the tune is next to nothing; the performer—with just that temperament, that range, that attack, that line, that relation to the pulse of the rhythm—is next to everything. Of course a session can be recorded; that is the sense in which it can in principle be made permanent. But that session cannot be performed again; that is the sense in which it is evanescent. Succeeding performances of a play arise from the production, not independently from the play; succeeding sessions of a jazz group arise from the state and relations of the players, and if from a preceding performance, then as a comment on it. (Modern performance may negate such distinctions; it does not annihilate them.) I daresay the themes moving from Shakespeare to film are richer than the tunes of jazz. But the matter of life and death, of whether these themes actually survive in film, is a matter of whether they find natural transformations into the new medium, as in moving from life in the water to life in the air. The feature of the medium of film I have just emphasized suggests that acting for film is peculiarly related to the dimension of improvisation, that there is for film a natural dominance of improvisation over prediction, though of course each requires (its own form of) technique and preparation. (This dominance is a specification of a description I have given elsewhere concerning film's upheaval of certain emphases established in theater, namely, that for film there is in acting a natural ascendancy of actor over character. This matter of the film actor's individuality will come back.)

I was talking about the emphasis on the father-daughter relation in these dramas. The classical obligations of the father in romance are to provide his daughter's education and to protect her virginity. These obligations clearly go together; say they add up to suiting her for marriage. Prospero describes or enacts his faithfulness to these obligations toward Miranda with didactic explicitness. In comedies of remarriage the *fact* of virginity is evidently not what is at stake. Yet all the more, it seems to me, is the *concept* of virginity still at stake, or what the fact meant is at stake—something about the possession of chasteness or innocence, whatever one's physically determinable condition, and about whether one's valuable intactness, one's individual exclusiveness, has been well lost, that is, given over for something imaginably better, for the exclusiveness of a union. The overarching question of the comedies of remarriage is precisely the question of what constitutes a union, what makes these two into one, what binds, you may say what sanctifies in marriage. When is marriage an honorable estate? In raising this question these films imply not only that the church has lost its power over this authentication but that society as a whole cannot be granted it. In thus questioning the legitimacy of marriage, the question of the legitimacy of society is simultaneously raised, even allegorized.

The specific form authentication takes varies in the various films. All, however, invoke the continuing question of innocence, sometimes by asserting that innocence is not awarded once for all, but is always to be rewon (I take this to be a way of telling the story of *Adam's Rib*); sometimes by asking what it means to lose innocence, and even to ask how the burden of chastity can be put off, or anyway shared (*The Philadelphia Story* is about the mystery in putting aside what we may call psychological virginity, an issue of Blake's poetry). In "Leopards in Connecticut" (Chapter 3) I argue that *Bringing Up Baby* contains, even consists of, an extended allegory of this question of sharing the loss of chastity; but a fully hilarious consciousness of the problem occurs right in the first of these films, with *It Happened One Night*, where the mutual happiness in the loss of virginity, or the happiness of mutuality, is said to require nothing less than what authorized the tumbling of the walls of Jericho, trumpet and all.

These parodies are themselves further parodied in *The Lady Eve*, as befits its mode, in its use of the "slimy snake" as an object of terror to the woman, conscious and unconscious—an object from which she awakens screaming, saying she dreamed about it all night. We are being clunked on the head with an invitation to read this through Freud. But the very psychological obviousness of it serves the narrative as an equivalent, or avatar, of the issue of innocence. It demonstrates that sexuality is for this sophisticated and forceful woman still a problem. No doubt this pokes fun at the older problem of virginity; what used to be a matter of cosmic public importance is now a private matter of what we call emotional difficulty. We live in reduced circumstances. But the obviousness also, I think, pokes fun at our sophistication, when that goes with a claim that we have grown up from ancient superstition. If virginity was a superficial and even idolatrous interpretation of the problem of innocence, with what has our sophistication replaced this idol?

One consequence of our sophistication is that if we are to continue to provide ourselves with the pleasure of romantic comedies, with this imagination of happiness, we are going to require narratives that do not depend on the physics of virginity but rather upon the metaphysics of innocence. In practice this poses two narrative requirements: that we discover, or recover, romance within the arena of marriage itself; and that a pair be capable of discussing with interest not merely the promises of love (topics of courtship—described by Harry in *The Lady Eve* as "whatever it is young people talk about") but the facts of marriage, which the facts of life they have shared require them to talk about. Comedies of remarriage typically contain not merely philosophical discussions of marriage and of romance, but metaphysical discussions of

THE LADY EVE

We may take the world she has in her hand as images in her crystal ball, but however we take it we are informed that this film knows itself to have been written and directed and photographed and edited.

the concept that underlines both the classical problem of comedy and that of marriage, namely, the problem and the concept of identity—either in the form of what becomes of an individual, or of what has become of two individuals. On film this metaphysical issue is more explicitly conducted through the concept of difference—either the difference between men and women, or between innocence and experience, or between one person and another, or between one circumstance and another—all emblematized by the difference, hence the sameness, between a marriage and a remarriage.

We got into the topics of virginity and chastity or innocence in naming the father-magician's obligations. The second obligation was that of seeing to his daughter's education, and really we are already addressing this topic in registering the necessity for philosophical or metaphysical discussion in these film comedies, because the form the woman's education takes in them is her subjection to fits of lecturing by the men in her life. For some reason Katharine Hepburn seems to inspire her men with the most ungovernable wishes to lecture her. Four of them take turns at it in *The Philadelphia Story*, and throughout *Adam's Rib* Spencer Tracy is intermittently on the verge of haranguing her. His major speeches take the form of discourses, one of them presenting his theory of marriage as a legal contract. Rosalind Russell does not escape this fate even from the Cary Grant of *His Girl Friday*. In *The Lady Eve*, the man's tendency to lecture nobly is treated to an exposure of pompous self-ignorance so relentless that we must wonder how either party will ever recover from it. (The woman describes this exposure as teaching a lesson, the spirit of which is evidently revenge; earlier she had saved him from what he calls "a terrible lesson your father almost taught me," namely, about games of so-called chance. Or was the lesson about disobeying this woman? She expressed particular impatience with him, quite maternal impatience, in saying, "You promised me you would not play cards with Harry again.")

Comic resolutions depend upon an acquisition in time of self-knowledge; say this is a matter of learning who you are. In classical romance this may be accomplished by learning the true story of your birth, where you come from, which amounts to learning the identity of your parents. In comedies of remarriage it requires learning, or accepting, your sexual identity, the acknowledgment of desire. Both forms of discovery are in service of the authorization or authentication of what is called a marriage. The women of our films listen to their lectures because they know they need to learn something further about themselves, or rather to undergo some change, or creation, even if no one knows how the knowledge and change are to arrive. (It turns out not to be clear what the obligations are for suiting oneself for remarriage.) In *It Happened One Night*, *His Girl Friday*, *The Awful Truth*, and *The Philadelphia Story* the woman imagines solving the problem of desire, or imagines that the problem will take care of itself, by marrying the opposite of the man she took first—an action variously described as the forgoing of adventure, and choosing on the rebound, and the buying of an annuity. Even if this man is not in fact older than her former husband, he is a father or senex figure, who must be overcome in order for the happiness of a comic resolution to happen. What our films show is that in the world of film if the woman's real father exists, he is never on the side of this father figure but, on the contrary, actively supports the object of her true desire, that is, the man she is trying, and trying not, to leave.

If this acceptance by the father of the daughter's sexuality, which means of her separation or divorce from him, the achieving of her human equality with him, is part of the happiness of these women, of their high capacities for intelligence, wit, and freedom, it also invites a question about the limitations of these comedies, about what it is their laughter is seeking to cover. The question concerns the notable absence of the woman's mother in these comedies. (The apparent exceptions to this rule serve to prove it.) The mothers that do figure in them are, blatantly, the mothers of the senex figures, separation from whom would not be contemplated. No account of these comedies will be satisfactory that does not explain this absence, or avoidance. I offer three guesses about regions from which an explanation will have to be formed. Psychologically, or dramatically, the central relation of a mother and son has been the stuff of tragedy and melodrama rather than of comedy and romance. (Shaw's *Pygmalion*, explicitly about the creation of a woman, is a notable exception to this rule; here the hero and his mother are happy inspirations to one another. But no less notably, if the central man and woman of this play find their way together at the end, it is explicitly to occur without marriage and its special intimacies.) We seem to be telling ourselves that there is a closeness children may bear to the parent of the opposite sex which is enabling for a daughter but crippling for a son. Eve will say to Charles on the train, "I knew you would be both husband and father to me." She says it to deflate him for his insincerity and hypocrisy, but what she says is true, and it is the expression of a workable passion. Whereas no one would be apt to hope for happiness (given the options we still perceive) should a man say to his bride, "I knew you would be both wife and mother to me." Whether you take this as a biological or a historical destiny will depend on where you like your destinies from. Mythically, the absence of the mother continues the idea that the creation of the woman is the business of men; even, paradoxically, when the creation is that of the so-called new woman, the woman of equality. Here we seem to be telling ourselves that while there is, and is going to be, a new woman, as in the Renaissance there was a new man, nobody knows where she is to come from. The place she is to arrive is a mythological locale called America. Socially, it seems to me, the absence of the woman's mother in these films of the thirties betokens a guilt, or anyway, puzzlement, toward the generation of women preceding the generation of the central women of our films—the generation that won the right to vote without at the same time winning the issues in terms of which voting mattered enough. They compromised to the verge of forgetting themselves. Their legacy is that their daughters will not have to settle. This legacy may be exhilarating, but it is also threatening.

Complementing the inability to imagine a mother for the woman is the inability to envision children for her, to imagine her as a mother. The absence of children in these films is a universal feature of them. What is its point? One might take its immediate function to be that of purifying the discussion, or the possibility, of divorce, which would be swamped by the presence of children. But what this means, on my view of these comedies, is that the absence of children further purifies the discussion of marriage. The direct implication is that while marriage may remain the authorization for having children, children are not an authentication of marriage. (This is an explicit and fundamental consequence of Milton's great tract on Divorce, a document I take to have intimate implications in the comedy of remarriage, as will emerge. By the way, the only claim among related comedies I know that a child is justified apart from marriage, even apart from any stable relationship with a man, merely on the ground that you bore it and want it and can make it happy, occurs in Bergman's *Smiles of a Summer Night,* I suppose the last comedy to study remarriage.)

But the films of our genre are so emphatic in their avoidance of children for the central marriage that its point must be still more specific. In *His Girl Friday* the woman's choice to remarry is explicitly a decision to forgo children as well as to forgo the gaining of a mother-in-law. In *The Awful Truth,* what room there is for a child is amply occupied by a fox terrier. In *The Philadelphia Story,* Grant's life without Hepburn is said by him to be, or described as, one in which he might as well part with a boat he devoted a significant piece of his life to designing and building—named *True Love*—on the ground that it is only good for two people. He means that to mean that one person is one too few for it, but his words equally mean that three is one too many. In *Adam's Rib,* as the principal pair are preparing some leftovers for supper, having chosen to stay home alone on cook's night off, there is a knock at the door which they know to be their wearisome childlike neighbor from across the hall. Tracy says to Hepburn, "Now remember, there's just enough for two." (I have not included George Stevens's *Woman of the Year* in my central group of comedies of remarriage because I do not find it the equal of the six or seven I take as definitive. But it speaks radically to the present issue. In it Spencer Tracy takes a child back to an orphanage from which Hepburn had adopted it out of concern for her public image as a leading woman. It is equally to the point that the older woman in this film, said by Hepburn always to have been her ideal, is not her mother but marries her father late in life, in the course of this film, in a scene that enables the younger woman to try for a reconciliation of her own.)

I do not think we are being told that marriages as happy as the ones in these films promise to be are necessarily incompatible with children, that the forgoing of children is the necessary price of the romance of marriage. But we are at least being told that children, if they appear, must appear as intruders. Then one's obligation would be to make them welcome, to make room for them, to make them be at home, hence to transform one's idea of home, showing them that they are not responsible for their parents' happiness, nor for their parents' unhappiness. This strikes me as a very reasonable basis on which to work out a future.

(It is perhaps worth making explicit that only when a period of culture is reached in which contraception is sufficiently effective and there is sufficient authorization for employing it conscientiously is it pertinent to speak of marriage quite in this way. There was a time—perhaps lived climactically in the generation of the absent mothers—when for a woman prepared to demand the kind of autonomy demanded by the women in our films, chastity, or anyway the absence of intimacy with men, would have presented itself as autonomy's clearest guarantee. The issue then would have been whether to have a recognized sexual existence at all, and hence, if marriage requires a sexual existence, whether to marry. But then if such a woman as dominates our films does choose to marry, risking children; if, that is, she requires a marriage in which children *can* be made welcome; then she is looking for a household economy which can undergo this transformation without her being *abandoned* to motherhood. This all the more for her puts the issue of marriage before the issue of children. The question of pregnancy is surely one of the reasons that feminism is thought to lack a sense of fun. Yet each of the women of our films is who she is in part because of her sense of fun, a sense apart from which the man in her life does not wish to exist. The question becomes what the conditions are—and first the requirements upon the man—under which that sense of fun can be exercised. So the conditions of the comic become the question of our genre of comedy.)

The insistence of these films on the absence of children seems to me to say something more particular still. Almost without exception these films allow the principal pair to express the wish to be children again, or perhaps to be children together. In part this is a wish to make room for playfulness within the gravity of adulthood, in part it is a wish to be cared for first, and unconditionally (e.g., without sexual demands, though doubtless not without sexual favors). If it could be managed, it would turn the tables on time, making marriage the arena and the discovery of innocence. *Bringing Up Baby*, on my account, is the most elaborate working out of this wish, but the value of it is fully present, for example, in the repeated remark of *The Philadelphia Story* that the divorced pair "grew up together"; and when Spencer Tracy goes into his crying act at the close of *Adam's Rib* (and we already know that their private names for one another are Pinky and Pinkie), he means to be demonstrating a difference or sameness between men and women, but he is simultaneously showing that he feels free to act like a child around this not obviously maternal woman. *The Awful Truth* ends with the pair dressed up in clothes too big for them, then being impersonated by two figurines doing a childish jig and disappearing together into a clock that might as well be a playhouse. This is in turn a further working out of the woman's having in the previous sequence put on a song-and-dance act in which she at once impersonates a low-class nightclub performer and pretends to be the man's sister, thus staking a final claim to have known him intimately forever.

The form taken by the search for childhood and innocence in *The Lady Eve* is given in that fantasy or romance the man tells the woman with its moral that he feels that they have known one another all their lives and hence that he has always loved her, by which he says he means that he has never loved anyone else. His attempt to repeat this story and to draw this moral again in Connecticut with Eve presents the most difficult moment of this comedy, the moment at which, as I put it earlier, their behavior toward one another appears unforgivable, hence the moment at which we may doubt most completely that a happy end for them can be arrived at. Some such moment must be faced in any good comic narrative; Sturges carries the moment to virtuosic heights in this film. And the question we have known awaits us is whether he succeeds in bringing the consequences safely to earth, or in blowing them sky high, in any case whether the film arrives at something we will call happiness for each of this pair and whether we are happy to see them arrive there.

But just what is the difficulty of this most difficult moment? Presumably that in repeating his romantic vision to Eve the man loses all claim to sincerity, which was really all that has recommended him to our attention. His story was hard enough to listen to the first time, when he told it to Jean, but we went with it because the woman's belief ratified it for us. On his repetition of it we do not know whether to be embarrassed more for him or for ourselves in being asked to witness this awful exposure. But how is his insincerity exposed? It is exposed only on the condition that we take him not to know or believe Eve and Jean to be one and the same woman. But must we so take him? I do not, of course, claim that he does know or believe that they are the same, that he is having to do with just one woman. But we have had continued evidence that he is in a trance (his word for this is "cockeyed"); and the fact of the matter is that he *is* saying his words to the same woman. What he says to Jean at the end is hard to deny: "It would never have happened except she looked so exactly like you." Furthermore, the comic falls the man keeps taking are more Freudian clunks on the head to tell us—as in the case of her reaction to his snake—that genuine feeling has been aroused, and moreover the *same* feeling that had been aroused by the woman on the boat whom he encountered by falling and who will once more enter his recognition through that same route of access. So his inner state as well as his external senses tell him that she is the same person. (So maybe the horse stands not merely for a horse laugh but also for the man's own natural instincts, but baffled by his efforts at domestication.) His intellectual denial of sameness accordingly lets him spiritually carve her in half, taking the good without the bad, the lady without the woman, the ideal without the reality, the richer without the poorer. He will be punished for this.

If we understand his perceptions and his feelings to be the same now as then, then we must understand ourselves to be embarrassed not by the openness of his insincerity but by the helplessness of his sincerity. He desperately wishes to say these words of romantic innocence to just this woman, even as she desperately wished to hear them. (This was a piece of her education.) Yet knowing this she feeds him with the fruit of the tree of stupidity. For this she will be punished.

Note the confluence of conventions Sturges activates in making up his story about identical twins. He gets the narrative and psychological complexities of early romantic comedy, with its workings out of mis-identifications and climaxes of recognition, together with a succinct declaration of the nature of film by way of showing its distinction from theater. For the stage, a convention allows two people dramatized as identical twins to be treated as though they cannot be told apart. For the screen, where two characters can be played by one person, and even shown side by side (a fact enjoyed in films from *Dr. Jekyll and Mr. Hyde* to *The Prisoner of Zenda*), a comparable convention allows a person to be treated as though he or she *can* be told, so to speak, apart from himself or herself, even where—and here Sturges rubs it in—she looks no dif-ferent from one role to the other. If we had taken Charles (or to the ex-tent that we take him) simply to believe that Eve was not the same woman as Jean, then (to that extent) we had been gulled as he had been—by the same story of romance; or anyway gulled at one remove from that story—by the film that suggests that he could simply believe such a story. (There are theories that believe so too, that assume that we do not know the difference between projections of things and real things and that therefore projections of reality are "illusions" of it.) How could we have believed this?

You might wish to give some further psychological explanation of the man here, but that would be to compete with him on his own level, for he has what he calls a piece of "psychology" that explains away to himself Eve's strategy. I think the ambiguity about whether he does or does not believe in her difference from herself is as fixed for us as it is for him. What it is fixed by is the photograph Hopsy is shown in order to reveal to him the (criminal) identity of Jean, along with Harry and Gerald. Hopsy learns this identity not from the photograph itself but from reading the caption printed on its back. The information con-tained in the caption is, of course, not news to us; what is news for us is the photograph itself. As it fills the screen, slightly inflected so as

They and their reflections are visible together to us, showing us that while these two can view the two worlds they move between, the one world from the conning perspective of the other, they may not occupy either wholly, or not at the same time (as with a thing and its filmed projection).

clearly to resist coincidence with the photographic field of the moving film images, what we are shown, and are meant to recognize that we are being shown, is a photograph of Barbara Stanwyck, Charles Coburn, and Melville Cooper. Or at the very least or most we are shown a photograph of Barbara Stanwyck as Jean Harrington, of Charles Coburn as Harry, and of Melville Cooper as Gerald. (It would be just like Sturges were the object we are shown to be, what it seems to be, a production still from the set of this film.) What this presenting of the photograph means to me is that we have a problem of identification isomorphic with this man's problem, one which lets his deluded or self-manufactured problem get a foothold with us, one which associates us with him in the position of gull. The relation between Eve and Jean is not an issue for us, but the nature of the relation of both Eve and Jean to Barbara

Stanwyck, or to some real woman called Barbara Stanwyck, is an issue for us—an issue in viewing films generally, but declared, acknowledged as an issue in this film by the way it situates the issue of identity.

It is a leading thought of mine about the film comedies of remarriage that they each have a way of acknowledging this issue, of harping on the identity of the real women cast in each of these films, and each by way of some doubling or splitting of her projected presence. I have already mentioned Irene Dunne's scene of impersonation in *The Awful Truth;* this bears comparison with Katharine Hepburn's gun moll routine in *Bringing Up Baby,* which refers back to it (by using the name "Jerry the Nipper"). From *It Happened One Night* through *The Philadelphia Story* to *Adam's Rib,* this splitting is investigated as one between the public and the private, where the public is typically symbolized by the presence of newspapers (or a news magazine)—a major iconographical or allegorical item in virtually every one of the films of our genre. It seems that film, in contrast to the publicity of newspapers, symbolizes the realm of privacy. In *Adam's Rib* this symbolism is most explicitly worked out as a split or doubling between what happens during the day and what happens at night, which amounts to a split or doubling between reality and something else, call it dreaming. The idea of the privacy of film is both confirmed and denied in *Adam's Rib,* say it is puzzled, by the showing on the first night of a home movie. (In another of George Stevens's films adjacent to our genre, *Talk of the Town,* the mode in which a copy of a newspaper is presented in order to reveal a hero's identity at the same time reveals newspapers to be things full of borscht. Again, by the way, this moment in which a front-page photo of a wanted man is the object of concern to two men and a woman about to have a meal together must be a reference to a moment in Hitchcock's *39 Steps.* We have here, I believe, one genre claiming its relationship to another.)

From the first of the major films of remarriage, *It Happened One Night,* the genre is in possession of the knowledge that the split or doubling is between civilization and eros. Newspapers are a medium of scandal, but what they mean by erotic scandal consists of triangles, crimes of passion, sensational marriages, and ugly divorces. What our films suggest is that the scandal is love itself, true love; and that while it is the nature of the erotic to form a stumbling block to a reasonable, civilized existence, call it the political, human happiness nevertheless goes on demanding satisfaction in both realms. This is in effect the terrible lesson Jean/Eve teaches Hopsy/Charles. When she vows to her father that she is going to be everything the man wants her to be, she means it as a blessing to them both. When she is treated to his treacherous lack of trust, or his overtrust in the wrong thing, the public thing, she turns the blessing into a curse. As if to say: Even after you know our passion for one another, and our fun together, you are still a sucker for romance and cannot acknowledge that passion may have a past of flesh and blood; very well, I'll show you the reality of your ideal; I'll give you a new perspective in Connecticut; I'll turn the night into an endless day for you. You refused to believe in me earlier, now I'll give you something you will feel compelled to believe; you thought you believed the worst about me before, here is something you will find worse. She is gambling, carrying out her instruction the night of their honeymoon on the train, that he will take the bait that makes the taker mad. Had he found a sense of humor to outlast his credulity and her anger, he would be able to charge her with stalling on her wedding night by putting up a barrier, between her and her husband, of a thousand and one bawdy tales. The possibility that she is stalling further compromises the purity of the lesson she thinks she is teaching, makes it even funnier and, if possible, even uglier.

It is not news for men to try, as Thoreau puts it, to walk in the direction of their dreams, to join the thoughts of day and night, of the public and the private, to pursue happiness. Nor is it news that this will require a revolution, of the social or of the individual constitution, or both. What is news is the acknowledgment that a woman might attempt this direction, even that a man and a woman might try it together and call *that* the conjugal. (It is roughly what Emerson did call that; but then, as you would expect, he did not expect to find it between real men and women.) For this we require a new creation of woman, call it a creation of the new woman; and what the problems of identification broached in these films seem to my mind to suggest is that this creation is a metaphysical enterprise, exacting a reconception of the world. How could it not? It is a new step in the creation of the human. The happiness in these comedies is honorable because they raise the right issues; they end in undermining and in madcap and in headaches because there is, as yet at least, no envisioned settlement for these issues.

CONS AND PROS

How does the film at hand end? How can any happiness at all be found in this revenge comedy?

Before drawing to its, and closer to our, conclusion, I note the most daring declaration of this film's awareness of itself, of its existence as a film. This comes by way of its virtual identification of the images seen on the screen with the images seen in a mirror. One plausible understanding of our view as Jean holds her hand mirror up to nature—or to society—and looks surreptitiously at what is behind her is that we are looking through the viewfinder of a camera. In that case this film is claiming that the objects it presents to us have as much independent physical reality as the objects reflected in a mirror, namely, full independent physical reality. Their psychological independence is a further matter, however, since we are shown Jean creating their inner lives for us, putting words into their mouths ("Haven't we met some place before? Aren't you the Herman Fishman I went to the Louisville Manual Training School with? You aren't?"), and blocking their movements for them ("Look a little to your left, bookworm. A little further. There!"), and evaluating their performances ("Holy smoke, the dropped kerchief!"). We may take the world she has in her hand as images in her crystal ball, but however we take it we are informed that this film knows itself to have been written and directed and photographed and edited. (Each of our films shows its possession of this knowledge of itself. *The Lady Eve* merely insists upon it most persistently.) That the woman is some kind of stand-in for the role of director fits our understanding that the man, the sucker, is a stand-in for the role of audience. As this surrogate she informs us openly that the attitude the film begins with is one of cynicism or skepticism, earned by brilliance, and that she is fully capable of being thus open and yet tripping us up so that we are brought from our privacy onto her ground, where her control of us will be all but complete. Frye notes that the inclusion of some event particularly hard to believe is a common feature of Shakespeare's comedy, as if placed there to exact the greatest effort from his dramatic powers and from his audience's imagination. And it is well recognized that the final two of Shakespeare's romances, *The Winter's Tale* and *The Tempest*, most clearly and repeatedly give consciousness to their own artifice, that they are plays with casts, as if no responsibility of art shall go unacknowledged. Then it may be in their awareness of themselves, their responsibility for themselves, that the films of remarriage most deeply declare, and earn, their allegiance to Shakespearean romance.

Further discussion of the significance of the phenomenon of mirroring in this film would have to take up the passage in which, the morning after Jean's triumph over her father at cards ("Know any more games, Harry?") and her ensuing receipt of Hopsy's proposal on the bow of the ship, she and her father begin an interview (as she is seated before the standing dressing mirrors in her stateroom and her father enters from the far door behind her, reflected in the mirror, and walks toward his reflection across the room to her) looking at one another in the mirror, speaking to each other's reflection first, communicating through the looking glass. What does this mean in this context? The mood is one of sober, even pained, sincerity and tenderness between them, as though the reflection of mirrors is not to be ceded to the realm of appearances but provides an access, or image, of self-reflection and thoughtfulness, of a due awareness of the world's awareness of you, hence of the other side of its reality to you. (The conjunction of mirrors with moments of sincerity, in a world of fashion and gossip, occurs notably in *Rules of the Game*.) In this interview the father warns his daughter that her admirer might not respond well to the truth about her and her father's lives: "You are going to tell him about us, aren't you?" She replies that a man who couldn't accept the truth wouldn't be much of a man. But all the time they and their reflections are visible together to us, showing us that while these two can view the two worlds they move between, the one world from the conning perspective of the other, they may not occupy either wholly, or not at the same time (as with a thing and its filmed projection). Here the camera especially ponders the meaning of a point of view, seeing these people and seeing what these people see at one and the same time, a feat they have to forgo in order to stand face to face.

One more preliminary to a conclusion, again having to do with fathers and movies and reflections that declare the presence (or distance) of real people. The opening of the shift to Connecticut discovers and follows Eugene Pallette walking down a long flight of period stairs as he sings, thoughtfully, "Come landlord fill the flowing bowl until it doth run over. For tonight we'll merry merry be, tomorrow we'll be sober." Criticism is being challenged to net in mere words the hilarity, the surrealism, the dream perfection of these juxtapositions; of its being just this human being doing just these things in just this setting. Here is Preston Sturges glorying in the modes of conjunction specific to film, and some specific to Hollywood, and indeed to America, making sure that we know that he knows what he is doing. The pivot of these conjunctions is that voice, declared by, of all things, singing, which declares the presence (by absence) of the only man who could possess it, call him Eugene Pallette, who brings with him, on that Tudorish staircase singing that Elizabethanish ditty, the world of Robin Hood in which he was (or perhaps is) Friar Tuck. (Melville Cooper was the Sherriff of Nottingham in the same production.) The existence of this man in that part no more and no less proves the irresponsibility and resourcefulness of Hollywood than the presence of Tudor mansions just north of New York City proves or disproves the irresponsibility and resourcefulness of American captains of business (though in both cases these presences bespeak a particular set of fantasies). By the time this ale merchant finishes his drinking song and his descent into the world, answers a telephone from which he learns that there is to be a party at his house that night, hangs up the receiver and responds by delivering an observation—"Nut house"—to no one in particular, casting a glance at his surroundings offscreen, we can sense that he is speaking for Sturges and that what he is looking at offscreen is a Hollywood sound stage. This memorable establishment of the hero's father as a character in possession of an inner life of independent judgment prepares him for a decisive function in the conclusion to be drawn by this film.

Now, how can this woman accept back her trusting/untrusting man, after what she has done to him? How can she hope for happiness with him, who at the end still does not know what has happened to him, hope that with him all will be well that ends well? She had said early on that he's touched something in her heart, and later on she confesses this awful yen for him. This combination of tenderness and sensuality is just what the doctor ordered for grown-up love in his *Three Essays on the Theory of Sexuality*. This text also contains, near its conclusion, a sentence that may stand as the motto for the entire genre of remarriage: "The finding of an object is in fact the refinding of it." But how does this woman work her way back to it? No doubt the man's very innocence, the completeness with which this mug appreciates her, the fervor as well as the sappy deliberateness with which he twice appeals to her to find an innocent past together, the very fact that he is what her father calls "as fine a specimen of the sucker sapiens as it has been my fortune to see"; no doubt all this, from being an object of her kidding, and of her scorn, finally elicits again her response in kind. And my question is, how?

I take the answer to be given in the man's father's appreciation of her (and the feeling is mutual) as he conveys his son's refusal to meet her sole condition for agreeing to divorce, that he come to her and ask her to be free. Here is a further thematic coincidence with *The Tempest*. (And what does "ask me to be free" mean? Ask me to let him go?; or, Ask me to let myself be free?) The father tells her he thought it was a pretty fair offer and says he thinks she is a sucker to make it. The father's lawyers are aghast at this businessman's sudden artlessness. Harry and Gerald are aghast at this metamorphosis of artist into sucker. She has become what the man is, a member of his species, the sucker sapiens, the wise fool; she has found what Katharine Hepburn at the end of *The Philadelphia Story* calls a human being; she has created herself, turned herself, not without some help, into a woman. She has done it by laying aside her art, call it her artifice; and in her long and passionate declaration to the man as she shuts them behind her cabin door, she virtually repeats his old story back to him, with the ending: "Don't you know I've waited all my life for you, you mug?"—thus confessing herself to be a mug. This concludes her education.

Mug is almost the last word we hear in the film, as it is one of the first, when her father responds to one of her professional questions by saying, with unquestionable wisdom, "A mug is a mug in everything." Her answer at the end of the film is, in effect: If to be at one with humanity is to be a mug, then as E. M. Forster almost put it, better a mug of the confidence game than a mug of the lack of confidence game, a mug of magic, of exemption. I should, of course, like to say that what she gets in return is another magic, not of control, but of reciprocity. But then you would think me a romantic.

But the word *mug* is not quite the last. The last is reserved for the character actually named a mug by the author of the film, anyway named a diminutive or a diminished mug: Mugsy (William Demarest). He has been remorselessly present from the beginning, but at the last possible moment he is expelled (fictionally, not cinematically, not in the same way). His provenance is clear enough. He is the melancholy that comedy is meant to overcome, the mood Frye notes as forming the opening of at least five of Shakespeare's comedies. This film further specifies this mood as the creature of suspicion and literalness. I think of him privately as a certain kind of philosophical critic, almost the thing Iago describes himself to be—"nothing if not critical." And faced with romantic flights of fantasy, with interpretations of feeling and conduct that make up dreamworlds of eternal and innocent love, who is there who will deny the truth of what Mugsy—the spirit of negativity—says?: Positively the same dame.

2

KNOWLEDGE AS TRANSGRESSION

It Happened One Night

Clark Gable is being parental, and he's so good at it that you don't know whether to consider that the paternal or the maternal side of his character predominates.

PHILOSOPHY is a subject, as Thoreau says about the subject of economy in the opening chapter of *Walden* (and by economy Thoreau means something very like what we may mean by philosophy), "which admits of being treated with levity, but it cannot so be disposed of." I hope it will be clear to the reader that, and why, I am not at pains always to maintain that philosophical issues raised by these films, and by film as such, can at all times be treated with a light heart—issues such as the search for identity discussed in the preceding chapter, or, in the present chapter, the acceptance of finitude. I gave reasons in the Introduction for my refusing to disguise, even for my wishing somewhat to court, an outrageousness in the subjects of film and of philosophy, especially at what I think of as their mutual frontiers. I introduced those reasons with the warning that the transgression of these frontiers would take its most extended form in the opening pages of Chapter 2, the reading of *It Happened One Night*, and I pause here, at the entrance to those pages, both to reiterate this warning and to make explicit the author's claim that the reader will indeed throughout, despite certain appearances, be reading the same book. Positively.

IF IT IS INEVITABLE that the human conceive itself in opposition to God; and as debarred from a knowledge of the world as it is in itself; and as chained away, incomprehensibly, maddeningly, from the possibility of a happy world, a peaceable kingdom; then it is inevitable that the human conceive itself as limited. But what is it to conceive this? Let us say it is to take ourselves as finite. Would this be something positive or something negative, something lacking? In either case, it portrays being human as being inherently subject to the fate of transgression, to commandments and prohibitions that are to be obeyed and that therefore *can* be disobeyed.

I have recently published a book in which philosophical skepticism is cast as a wish to transgress the naturalness of human speech. But skepticism is also described as a peculiarly human prerogative. My subject here shifts the wish to transgression from what might be called the natural to the social plane. Two of the fundamental human properties that human societies have been most anxious to limit are the capacity to relate oneself to the world by knowledge and the capacity to relate oneself to others by marriage. We seem to understand these capacities for relation as constitutive of what we understand by human society, since we attribute to them, if unchecked, the power to destroy the social realm.

If we do not equate human knowledge with the results of science but understand it as the capacity to put one's experience and the world into words, to use language, then the will to knowledge and the will to marriage may be seen to require analogous limitations in order to perform their work of social constitution, limitations that combat their tendencies to privacy or their fantasies of privacy. Concerning marriage, I am invoking Lévi-Strauss's understanding of the barrier to incest as the force necessary to compel that reciprocity and exchange apart from which separate human families cannot create the realm of the social, the public. I shall not claim that this understanding is as clear as one would like, and especially not that it is, as it seems to take itself to be, an alternative to Freud's psychological account. In particular, if there is a *horror* of incest, I do not see that Lévi-Strauss's compulsion to society reaches it. My intuition is merely that any better view of these matters, and of their connection, will have to take up this one. Concerning knowledge or language or naming, I am invoking Wittgenstein's construction and destruction of the possibility of a private language as revealing the barrier to narcissism, facing us with that reciprocity and exchange apart from which separate human individuals cannot acquire the force so much as to name themselves, to create the realm of the private.

This region of issues may seem as abstract and distant from our everyday experience as any theories of anthropology and of philosophy, but they are as close to us as what I am claiming to be a genre comprising the best of the Hollywood comedies since the advent of the talkie—or so I will try to show here in reading Frank Capra's *It Happened One Night*. Before turning to that film, I wish to map out a little the abstractness and the apparent distance of the idea of our lives as shaped by certain reigning intellectual and social barriers. I wish to call to mind certain pictures of the human projects of knowledge and of community blocked out most memorably in Kant's philosophizing, for Kant is the first figure likely to occur to a philosopher thinking about the subject of limits in human knowing.

The empiricists Locke and Hume also insist upon something each *felt* as limitations of reason. Locke's introduction to his *Essay Concerning Human Understanding* explains the motive for his investigation as one of determining, before any particular investigations of nature, or of God, what we can hope to know, what we are humanly equipped to know, to save ourselves fruitless ventures into matters exceeding the limits of our understanding. And in the *Dialogues Concerning Natural Religion*, Hume, through Philo, speaks of our understanding or experience as limited in extent and duration. Both recommend a modesty, or humility, in the exercise of our powers of understanding, suggesting a prudent limitation of our aspirations that will acccord better with these powers. But both Locke and Hume rather suggest that if our powers of understanding were enlarged, we would be in a position to know what we cannot at present know. What Kant undertakes to show is that our present position is in a way worse than that suggests and in a way better than that suggests. It is worse because although with "increased" range of experience and greater powers of understanding (and it is important that it is quite unclear what this would mean—new sensations? new concepts? new laws?) we would, of course, know *something* more than we know now, we would still never know enough, or not the right sort of thing, to satisfy the philosopher's dream of perfect knowledge. Philo had said that "our experience, so limited in extent and duration, can never provide us with a significant conjecture concerning the whole of things." It is as if Kant were saying: This formulation puts our problem wrongly from the beginning, it is a false picture of the faculty of knowledge altogether; for "the whole of things" *cannot* be known by human creatures, not because we are limited in the extent of our experience, but, as we might say, because we are limited *to* experience, however extensive. Put it this way: to know the world as a whole, or the world as it is in itself, would require us to have God's knowledge, to know the world the way we more or less picture God to know the world, with every event and all its possibilities directly present. And this simultaneous, immediate intuition of the world is not merely beyond us in fact or in extent; it is not a matter of having more of something we now have a little of. It is beyond us in principle; human knowledge is not like that. First, because all our knowledge, being a function of experience, is *sequential*; it takes place in time (in history, Hegel will say). Second, because the sequences of experience are categorized in definite ways—in terms of a definite notion of what an object is, of what a cause is—and there is no way to know whether these categories of the understanding are ultimately true of things. All we can say is, they are ours, it is our world. But our position is also *better* than Locke and Hume suggest. Because the discovery of our necessary limitations, our subjection to our experience and our categories, is one of human reason's greatest discoveries, it is the great discovery of reason about itself. The very facts that from one point of view are to us limitations of human knowledge are from another the necessary conditions of knowledge as such; and, therefore, in knowing these conditions once for all, we know once for all the general conditions or specifications or features anything must have in order to become an object of knowledge for us at all. And to know this is a traumatic increase of human knowledge.

Hence, to Hume's skeptical discovery that we cannot know, for example, that causation holds sway in nature, or that if it does, it will continue, that we know simply that certain experiences in fact follow other experiences, but that there is no necessity here, Kant's counter is this: the rule of causation and the other rules associated with the categories of the understanding are necessary in the sense that if they did not hold of the things of the world, there would be nothing to understand as a world of things. The categories of the understanding are interpreted by Kant as rules or laws that are imposed upon the material provided to thought by the system of the senses.

I note at once that we already have here a source of dissatisfaction with the more or less unrefined idea—to be found, it seems to me, in Lévi-Strauss's *Elementary Structures of Kinship*—that culture as distinct from nature is a realm of rules. What we understand as nature is also a realm of rules, that is, a realm, a world. For culture to be achieved, human beings, or rather rational creatures, must act, as Kant puts it, not merely in accordance with law but in accordance with the concept of law. We must understand ourselves as subject to law, and as the bringers of law. In both our participation in the world of nature and in the world of culture, reason acts as lawgiver, imposing order on our otherwise arbitrary and inconstant sensuous endowment. In each case reason provides our motions and our motives with that necessity and universality apart from which we would have no access to the objective, no idea of a world. (This hardly settles the Kantian question whether Reason is a unity.) In each case two worlds are discovered, and in each case there is between us and one of these worlds a barrier, establishing the condition of the world we normally live in as limited—call this the world of experience or of knowledge, or the public world, the shared world.

Kant's vision seems to tap various sources of our idea of our finitude, from the prohibitions in the Garden of Eden to the overreachers of tragedy to our various vague senses of our unbridgeable distance from nature and from others. But his idea of a limitation on human knowledge fixed by the fixed nature of the human being has caused as much perplexity as it has conviction. Is his idea of limitation (whatever the particular limitations he draws) the necessary consequence of his philosophizing, or is it to be accepted rather as a Wittgensteinian "picture," some sort of rigid fantasy of how things must be, itself in need of deciphering?

Some twenty years ago I remarked that Wittgenstein's criticisms of metaphysical speculation in *Philosophical Investigations* are a continuation of Kant's critique of metaphysical speculation,* specifically on three counts: in the emphasis on the "possibilities of phenomena"; in the provision of philosophical diagnosis of philosophical failure; and in the appeal to the idea of limits in this diagnosis. By now the idea of a general relation between Kant and the later Wittgenstein seems to be easily accepted, so perhaps it is worth now specifying some differences. First, there is for Wittgenstein no final systematic form of philosophy in the face of which metaphysical speculation can be brought to a halt. Nor can you always tell by looking, so to speak (looking, perhaps, at the

* "The Availability of Wittgenstein's Later Philosophy," in *Must We Mean What We Say?*, p. 65.

topic of a remark), whether a stretch of thought is metaphysically speculative. You have to try it out. The temptation to metaphysics becomes in Wittgenstein a will to emptiness, to thoughtless thought; and this is something that has to be resisted again and again, because the temptation to speculation, however empty, is as natural to the human creature as its criticism is. Second, the idea that what happens to the philosophic mind when it attempts speculation beyond its means is that it transgresses something we want to call limits, is an idea that cannot as it stands constitute a serious term of criticism for Wittgenstein but must remain merely a "picture," however significant. Kant, however, really does take the mind as *confined* in what it can know, takes it that there are things beyond the things we know, or something systematic about the things we know, that we cannot know, a realm of things-in-themselves, noumenal, open to reason, not phenomenal, not presentable. When Wittgenstein speaks of "bumps that the understanding has got by running its head up against the limits of language," the very obviousness of figurative language here works to suggest that thought is not confined *by* language (and its categories) but confined *to* language. And then we have to go on to ask, testing the picture: Is this really confinement? Is our freedom checked? From what are we withheld? I do not of course deny the presence of a *sense* of confinement here. On the contrary, I imagine that good answers to the questions I am asking will provide useful expressions of this sense, expressions concerning how we are to apprehend the picture of a metaphysical limit or barrier in our relation to the world and to others.

There is a notable difference in the ways we might grasp Kant's idea, as I put it, of reason as lawgiver, as world-creating, in the realm of the natural and in the realm of the social. In our knowledge of nature we seem to have no choice over whether the laws of our reason apply to it or not. What alternative could there be to the knowledge of nature? Ignorance of nature? But since Kant's problem is not a matter of knowing certain facts or specific physical laws of nature, but rather a matter of establishing the possibility of knowing nature at all, establishing the world of things as such, then ignorance of nature would mean our ignorance of such a thing as a world at all. To choose such ignorance would be like trying to choose to be an animal or an insect. Even that is not radical enough, since animals know something, even a totality of somethings. It would perhaps be expressible as our trying to choose to become stone, purely material. In the case of our social life we *do* have a choice over whether the laws of the moral universe, "objective" moral laws, apply to us; which is to say, a choice over whether to apply them, as is implied in their presenting themselves to us as imperatives, matters, as it were, not fully natural to us. This is as we should expect. There is an alternative to moral goodness—moral evil. Moral evil is not merely a matter of falling short of the dictates of the moral law: our sensuous nature indicates to us that for all we know we always fall short. The matter is rather one of choosing evil, of choosing to thwart the very possibility of the moral life. Kant does not say much about this alternative, but I understand it in the following way. One inflection of the moral law is that its necessity and universality are to be viewed as holding in "the realm of ends," which may be thought of as the perfected human community. This realm is also a world "beyond" the world we inhabit, a noumenal realm, open to reason, standing to reason; but I am not fated to be debarred from it as I am from the realm of things-in-themselves, by my sensuous nature; for the perfected human community *can* be achieved, it may at last be experienced, it is in principle presentable. Yet, there is between me and this realm of reason also something that may present itself as a barrier—the fact that I cannot reach this realm *alone*.

Any teacher of philosophy will have some way of picturing our inaccessibility to the realm of things-in-themselves, say by a circle or a line, readily drawn on the blackboard, outside of which or below which our mental and sensuous faculties cannot penetrate. (Some teachers might draw the same ready diagram year after year for a lifetime, each time with more or less the same sense that our human fate is being inscribed or emblematized. I assume that some of these teachers will be on a good path, some on a bad, depending on their capacities for diagramming, that is, for allegorizing.) But if now I ask myself how I picture the barrier to the realm of ends, I find I draw a blank. Would a good picture be an outline of my body, as of the perimeter of my power? Or ought I rather to try imagining the collection of all persons apart from me, with whom I know I ought to be, but am not, in community? Or is the absence of a picturable barrier here due rather to my not attributing the limits of community to a set of circumstances (as, for example, the sensuous dimension of human nature) but to a condition of will, together with my not knowing how to picture the will? But what in particular about the will? If the eventual community of humanity is not merely something close to us that we are falling short of, but something closed to us, something debarred, then its nonexistence is due to our willing against it, to the presence of moral evil. This takes moral evil as the will to exempt oneself, to isolate oneself, from the human community. It is a choice of inhumanity, of monstrousness. Then our inability to picture ourselves as debarred from the social, or as debarring it, our drawing a blank here, may express a horror of this possibility, call it a horror of metaphysical privacy, as though picturelessness were a kind of namelessness. (This choice, or refusal to apply the moral law to ourselves, is not, I think, to be understood as the disobedience in which Paradise is lost. As creatures who have lost an immediate connection with the commandedness, we are all disobedient; our obedience is forced, it is imperative, we would exempt ourselves if we could. Thoreau's way of saying this is to describe us as hard of hearing. Raskolnikov is not merely disobeying the law in a given case, refusing to universalize his maxim and act for the sake of the law. He might be conceived as attempting to refuse the law as such, to act for the sake of immorality, to become, let us say, unjudgeable. He purifies our wish for inhumanity. Whereas our everyday human, impure disobedience creates not hell but a restive and populous earth.)

NOT KNOWING whether human knowledge and human community require the recognizing or the dismantling of limits; not knowing what it means that these limits are sometimes picturable as a barrier and sometimes not; not knowing whether we are more afraid of being isolated or of being absorbed by our knowledge and by society—these lines of ignorance are the background against which I wish to consider Frank Capra's *It Happened One Night* (1934). And most urgently, as may be guessed, I wish to ponder its central figure of the barrier-screen, I daresay the most famous blanket in the history of drama. I am not unaware that some of my readers—even those who would be willing to take up Kant and Capra seriously, or earnestly, in isolation from one another—will not fully credit the possibility that a comic barrier, hardly more than a prop in a traveling salesman joke, can invoke issues of metaphysical isolation and of the possibility of community—*must* invoke them if this film's comedy is to be understood. I still sometimes participate in this doubt, so it is still in part myself whose conviction I seek.

The blanket dividing the space, and falling between the beds, is the man's idea as the principal pair, for the first of three times we will know about, prepare to share a cabin in an auto camp. The woman is understandably skeptical: "That, I suppose, makes everything all right." He replies that he likes privacy when he retires, that prying eyes annoy him, and goes on at once to situate the blanket allegorically: "Behold the Walls of Jericho. Maybe not as thick as the ones Joshua blew down with his trumpet, but a lot safer. You see, I have no trumpet." Wise in the ways of Hollywood symbolism, as generally obvious as the raising and lowering of a flag, we could already predict that the action of the film will close with the walls tumbling down. But then let us be wise enough, if we care about this film, to care about the rigors of this symbolism. The question the narrative must ask itself is how to get them to tumble. That this is a question, and the kind of question it is, is declared late in the film when the second blanket is shown unceremoniously pulled down by the suspicious owners of this second auto camp. Of course it is easy to pull it down if you do not know what it is, or care. So an early requirement for its correct tumbling is that the pair come to share a fantasy of what is holding it up.

An immediate complication is insinuated concerning who must use the trumpet. As the man, the wall in place, their spaces ready for the night, prepares to undress, he says: "Do you mind joining the Israelites?"—that is, get over to your side of the blanket. Now anyone who knows enough to refer to the Walls of Jericho—say a Hollywood script writer—knows that the Israelites are the attacking force and that it is they who have the relevant trumpets. Thus the man is repeating his claim that he has no trumpet and is adding that whether the walls come down will depend on whether the right sounds issue from her side of the wall. You may think this is pushing popular biblical study too far, but while it may be most common for audiences to interpret the allegory so that Clark Gable is Joshua and at the end blows the trumpet, it should be considered that we do not *see* this and that, for all we are apprised of, we are free to imagine that it is the woman who is still invited to make the move and who gallantly accepts the invitation. (So why don't we exercise that freedom?) If the trumpet is the man's, than presumably the blanket-wall represents the woman's virginity, or perhaps her resistance, even conceivably her reserve. I shall not deny that these symbolisms are in train here, but I wish to leave it open to the film to provide us with some instruction about what, a third of the way through our century and for a couple of persons not exceedingly young, virginity and resistance and reserve consist in, what the problem is about them.

I guess I would not place such emphasis on the possible ambiguity concerning who blows the trumpet apart from my taking this film as one defining the genre of the comedy of remarriage; for it is an essential feature of that genre, as I conceive it, to leave ambiguous the question whether the man or the woman is the active or the passive partner, whether indeed active and passive are apt characterizations of the difference between male and female, or whether indeed we know satisfactorily how to think about the difference between male and female. This is why I said that this genre of film rather refuses the distinction between Old Comedy and New Comedy, in the former of which the woman is dominant, in the latter the man. This is also a reason I have also called the genre the comedy of equality. Before going further into the genre here, however, let us notice something else we know about the blanket-barrier from almost the first moment it is put up.

The woman has joined the Israelites, the man finishes changing into his pajamas and gets into his bed, the woman asks him to turn off the light, after which she begins hesitantly to undress. In one camera set-up we watch the blanket-screen with the man as it is rippled and intermittently dented by the soft movements of what we imagine as the woman changing into pajamas in cramped quarters. The thing that was to "make everything all right" by veiling something from sight turns out to inspire as significant an erotic reaction as the unveiled event would have done. Call this thing the elaboration or substitution of significance, call it the inspiration of significance, the beginning of a credit system. The barrier works, in short, as sexual censorship typically works, whether imposed from outside or from inside. It works—blocking a literal view of the figure, but receiving physical impressions from it, and activating our imagination of that real figure as we watch in the dark—as a movie screen works.

I cannot doubt that the most celebrated Hollywood film of 1934 knows that it is, among other things, parodying the most notorious event of the Hollywood film's political environment in 1934, the accep-

What this pair does together is less important than the fact that they do whatever it is together, that they know how to spend time together, even that they would rather waste time together than do anything else—except that no time they are together could be wasted.

tance of the motion picture Production Code—the film industry's effort, it said, to avoid external censorship by imposing an internal censorship.* (Some avoidance; some originality.) The question posed by the parody may be formulated this way: If the film screen works like a kind of censoring, elaborating the effect of what it covers, how will you censor *that?*

Now we must start asking specifically what there is between just these two people that just this mode of censoring or elaboration is constructed between them. And for this a further elaboration of certain features of the genre of remarriage comedies will help. In Chapter 1 I traced out an emphasis on the father-daughter relation that the comedy

* Robert Sklar's *Movie-Made America* (New York: Vintage Books, 1976) has a good account of this event and useful references.

of remarriage inherits from Shakespearean romance, and I put together with this the absence of the woman's mother. The father's dominating presence is handled most wittily in *The Lady Eve* and most oratorically in *The Philadelphia Story*; but it is given its most pervasive handling in *It Happened One Night*. The entire narrative can be seen as summarized in the first of the newspaper headlines that punctuate it: Ellie Andrews Escapes Father. And throughout her escapades with Clark Gable, Claudette Colbert is treated by him as a child, as his child, whose money he confiscates and then doles back on allowance, whom he mostly calls "Brat," and to whom he is forever delivering lectures on the proper way to do things, like piggyback or hitchhike. After his first lecture, on the proper method of dunking doughnuts, she even says, "Thanks, Professor," a title more memorably harped on in *The Philadelphia Story*. In the genre of remarriage the man's lecturing indicates that an essential goal of the narrative is the education of the woman, where her education turns out to mean her acknowledgment of her desire, and this in turn will be conceived of as her creation, her emergence, at any rate, as an autonomous human being. ("Somebody that's real," the man will say, half out of a dream-state, at the climax of the film, "somebody that's alive. They don't come that way any more.")

But perhaps I should justify including this film under the genre of remarriage at all, since while it is true that a late newspaper headline satisfyingly declares Ellen Andrews Remarries Today, what the film— or the newspaper—thinks it means is not that she is to marry the real object of her desire again. I might say that what a film, or any work, thinks it means—or what one might at first think it thinks it means—is not to be taken as final. I might, again, say that the matter of remarriage is only one of an open set of features shared by this genre of comedy and that the absence of even that feature may in a given instance be compensated for by the presence of other features. Most pointedly, here, a film that opens (virtually) with the following exchange between a daughter and her father—

ELLIE: Can't you get it through your head that King Westley and I are married? Definitely, legally, actually married. It's over. It's finished. There's not a thing you can do about it. I'm over twenty-one, and so is he.

ANDREWS: Would it interest you to know that while you've been on board, I've been making arrangements to have your marriage annulled?

IT HAPPENED ONE NIGHT

—by that fact alone has a claim in my book to be called a comedy of remarriage, because a central claim of mine about the genre is that it shifts emphasis away from the normal question of comedy, whether a young pair will get married, onto the question whether the pair will get and stay divorced, thus prompting philosophical discussions of the nature of marriage. We might accordingly say here that the issue of remarriage is present but displaced. (Is there a reason this film opens on a yacht, beyond the obvious economy in establishing a setting of luxury? A boat is a good place for a father to confine a daughter without brutality, as, say, by locking her in a tower; and a ship's captain is empowered to perform a marriage ceremony but not to grant a divorce—as if the latter had not the same urgency. Yet, at the end the father-captain will perform a kind of divorce ceremony [called buying someone off], and it will be declared to be urgent.)

The idea of displacement seems to me right as far as it goes, but it does not explain how the issue gets displaced onto just this pair, what it is about them that invites it. It feels at the end as if they are marrying again, and not merely because of the plain fact, significant as it is, that the wedding night is shown to be set in yet another auto camp—which thus repeats two of the three nights they have already spent together—but specifically because what we have been shown in the previous auto camps is something like their marriage. We know of course that they have not been legally, actually married, but we also know that those things do not always constitute marriage, and we may freely wonder what does. Our genre is meant to have us wonder. The opening exchange between daughter and father, leading up to the response about annulment just quoted, had gone as follows.

ELLIE: I'm not going to eat a thing until you let me off this boat.

ANDREWS: Aw, come now, Ellie. You know I'll have my way.

ELLIE: Not this time you won't. I'm already married to him.

ANDREWS: But you're never going to live under the same roof with him. Now I'll see to that.

As if living under the same roof were the consummation apart from which a marriage may be annulled. The night of the second auto camp, which succeeds the night in the hay followed by a whole day of walking and hitchhiking and giving a thief a black eye for his car, the pair achieves something like marital familiarity as they prepare for bed, a familiarity heightened by the fact that what they are discussing is the intimate topic of never seeing one another again, and by the surrealist matter-of-factness with which the man goes about the business of hanging the blanket-wall-screen before each matter-of-factly undresses for bed. They are living under the same roof.

This familiarity is prepared by a former one, equally if oppositely powerful, at the first auto camp, when their breakfast is interrupted by Ellie's father's private detectives and the pair pretend to be a working-class married couple. Or rather, since in taking the cabin together they were already pretending to be man and wife (or pretending so again, since they have *already* already pretended as much, to put the quietus on the loudmouth forcing his attentions on her on the bus), we might say they are giving a charade of marriage. While this also does not achieve marriage, it does achieve the earlier of the familiarities I mentioned, since it makes their pretense of marriage by contrast seem an almost natural estate. The form the marriage charade takes is yet more significant. The pair mean the routine to convince hardened, suspicious observers on the spot that they are a seasoned couple, and their knock-down proof is to bicker and scream at each other. This laugh over the misery of a squalid, routine marriage poses at the same time a puzzle over the almost incessant bickering the pair have engaged in on their own from the instant they meet and dispute a seat on the bus. As if there may be a bickering that is itself a mark, not of bliss exactly, but say of caring. As if a willingness for marriage entails a certain willingness for bickering. This strikes me as a little parable of philosophy, or of philosophical criticism.

The exchanges of comedy span the quarrels of romance and the tirades of matrimony, arguments of desire and of despair. So essential are these arguments to the genre of remarriage that it may be taken above all to pose the problem: What does a happy marriage *sound* like? Since the sound of argument, of wrangling, of verbal battle, is the characteristic sound of these comedies—as if the screen had hardly been able to wait to burst into speech—an essential criterion for membership in that small set of actors who are featured in these films is the ability to bear up under this assault of words, to give as good as you get, where what is good must always, however strong, maintain its good spirits, a test of intellectual as well as of spiritual stamina, of what you might call "ear."

IT HAPPENED ONE NIGHT

IN THIS, as in other respects, these comedies illustrate, or materialize, the view of marriage formulated in John Milton's eloquent *Doctrine and Discipline of Divorce,* which I might describe as a defense of marriage in the form of a defense of divorce. It will further my argument to insist on this a little by quoting a summary statement from chapter 2 of Milton's theological point of departure.

> And what [God's] chief end was of creating woman to be joined with man, his own instituting words declare, and are infallible to inform us what is marriage and what is no marriage, unless we can think them set there to no purpose: "It is not good," saith he, "that man should be alone. I will make him a helpmeet for him" (Genesis 2:18). From which words so plain, less cannot be concluded, nor is by any learned interpreter, than that in God's intention a meet and happy conversation is the chiefest and noblest end of marriage, for we find here no expression so necessarily implying carnal knowledge as this prevention of loneliness to the mind and spirit of man.

(An Existentialist may regard hell as other people, as in Sartre's *No Exit.* But for a sensibility such as Milton's, myself am hell.) A modern reader of this passage is apt to feel that Milton's meaning of conversation in marriage is too remote from what we mean by conversation to apply to the exchanges between the pairs of our comedies. But why? Because Milton means something more by conversation than just talk, because he means a mode of association, a form of life? We might say he means something more like our concept of intercourse, except that our word conversation explicitly, if less generally, also carries the sexual significance as well as the social (as in the legal phrase "criminal conversation"). Contrariwise, Milton does also mean talk, as in the phrase "mute and spiritless mate" from his chapter 3—or at the least he means articulate responsiveness, expressiveness.* More important, the films in question recapture the full weight of the concept of conversation, dem-

---

* Milton's taking the passage from Genesis as what I called his theological point of departure was common theological practice when he wrote. What I am assuming is uncommon in him is his appeal from that passage to the idea of conversation. For the common practice, with illuminating excerpts and helpful bibliography, see Edmund Leites, "The Duty to Desire: Love, Friendship, and Sexuality in some Puritan Theories of Marriage," in *Structures of Consciousness, Civilizational Designs,* ed. Vytautas Kavolis and Edmund Leites (Madison, N.J.: FDU Press, 1980).

onstrating why *our* word conversation means what it does, what talk means. In those films talking together is fully and plainly being together, a mode of association, a form of life, and I would like to say that in these films the central pair are learning to speak the same language. (Of course this is learning to hear, to listen, as if loving and honoring were already grasped in the correct or relevant mode of obeying, the traditional promise of which is marriage.) That the language is private or personal or contains privacy is suggested by its being made explicit, for example, that they alone know what "the Walls of Jericho" means (though we are privy to its meaning). Their extravagant expressiveness with one another is part of the exhilaration in these films, an experience in turn possible only on the basis of our conviction that each of them is capable of, even craves, privacy, the pleasure of their own company. We understand something like the capacity for their pleasure, under threat by the erotic charge between them, by the demand to forgo one autonomy for another (or one idea of autonomy for another), to cause their hot hostility toward one another.

What this pair does together is less important than the fact that they do whatever it is together, that they know how to spend time together, even that they would rather waste time together than do anything else—except that no time they are together could be wasted. Here is a reason that these relationships strike us as having the quality of friendship, a further factor in their exhilaration for us. Spending time together is not all there is of human life, but it is no less important than the question whether we are to lead this life alone.

In stressing the ascendancy of being together over doing something together, the problem of these narratives requires a setting, as said, in which the pair have the leisure to be together, to waste time together. A natural setting is accordingly one of luxury, or as Frye puts it concerning romances generally, a setting for snobs. At least the settings require central characters whose work can be postponed without fear of its loss, or in which the work is precisely the following of events to their conclusions (rather than the gridding of days into, say, the hours of nine to five), as, for example, the work of a newspaper reporter. *Bringing Up Baby* presents the purest example of a relationship in which the pair do next to nothing practical throughout our knowledge of them; what they do is something like play games; you could almost say they merely have fun together, except that it takes the entire course of the film for the man to come to the essential insight about himself that he was throughout having fun. I would like to say that they achieve purposefulness without purpose. It is because of this purity of action, I believe, that people sometimes find *Bringing Up Baby* the hardest of these films to take.

But is it true of *It Happened One Night* that the pair are really wasting time together? After all, what she is doing is running away, and what he is doing is his job, getting a scoop. I do not wish to answer this merely by saying that it turns out that they are not doing those things really. They may have been doing them and then changed their minds at the last moment. I do not even wish to answer merely that they begin changing their minds almost from the first, as a result, say, of the marriage charade. I also wish to ask whether one can accept any such description of their work—as escaping and as scooping—as dictating the way they behave toward each other, his fathering and lecturing her, her playfulness and her achievement of humility. I will later dwell on the man's confusion toward the end in leaving the woman asleep and going to sell his scoop. Let me call his confusion, by way of anticipation, a matter of trying to sell the fiction that he has just at the end changed his mind, that this happened just last night, just one night, instead of long ago, and continuously—I mean, sell this to himself.

The recent theme of ambivalence, of the pair's revolving positive and negative charges, together with the theme of activeness and passiveness touched upon before, must also require placement for us in certain texts of Freud—for example, in this juxtaposition from *Civilization and Its Discontents*, footnoting some factors that contribute to civilization's dampening of "the importance of sexuality as a source of pleasurable sensations, that is, as a means of fulfilling the purpose of life":

> If we assume it to be a fact that each individual has both male and female desires which need satisfaction in his [or her?] sexual life, we shall be prepared for the possibility that these needs will not both be gratified on the same object, and that they will interfere with each other, if they cannot be kept apart so that each impulse flows into a special channel suited for it. Another difficulty arises from the circumstance that so often a measure of direct aggressiveness is coupled with an erotic relationship, over and above its inherent sadistic components. The love-object does not always view these complications with the degree of understanding and tolerance

manifested by the peasant woman who complained that her husband did
not love her any more, because he had not beaten her for a week.*

This last turn of masculine humor, by the way, is taken up precisely as
It Happened One Night is winding up. The woman's father asks the man
whether he loves his daughter (having already seen it proved that he
does not love his daughter's money). One of the man's responses is
this: "What she needs is a guy that'd take a sock at her once a day
whether it was coming to her or not." The father smiles, understanding
this as a trustworthy expression of true love; he has found the man after
his own heart; someone, as he will put it to his daughter, to make an old
man happy.

Before deciding that we have here one more example of a Stone Age
he-man, portrayed by a royal member of the species of the Hollywood he-man, and licensed by the director whose sentimentality is just
the other face of the fixated split between the masculine and the feminine, let us continue our supposition that this film has something to
teach us about our pursuits of happiness.

I think we should be surprised, given a certain conventional view of
what Clark Gable is, to find him capable of the sharp and playful conversation our genre requires. Then we must be prepared for astonish-
ment if we are to perceive the region of this he-man temperament
called upon in the breakfast sequence that prepares the staging of the
marriage charade that first morning after the first night in an auto camp.
He has walked into the cabin with a full grocery bag, tossed over to her,
still in bed, a package that turns out to contain a toothbrush, gruffly ordered her to get out of bed and to take a shower and get dressed because breakfast will be ready in no time, given her his robe and slippers
to wear, and he turns to preparing a meal; he has already, it emerges,
pressed her dress. Clark Gable is being parental, and he's so good at it
that you don't know whether to consider that the paternal or the maternal side of his character predominates. One would like to say, in
view of the representations he has made to be married to her, that he is
being a husband who understands that role as a classical commitment
to being both father and mother to the woman, except that his behavior
so far seems produced not by a response to her but by some conception
he has of himself. The revelation of his nurturant side is matched by

* (London: Hogarth Press, 1930), p. 77.

the revelation of her appreciation of it, neither resenting it nor taking it for granted.

A MAJOR THEMATIC DEVELOPMENT is under way, based on food. Here are the opening words of the film, preceding the initial interview quoted earlier between father and daughter.

ANDREWS: On a hunger strike, eh? How long has this been going on?

CAPTAIN: She hasn't had anything yesterday or today.

ANDREWS: Send her meals up to her regularly?

CAPTAIN: Yes, sir.

ANDREWS: Well, why don't you jam it down her throat?

CAPTAIN: Well, it's not as simple as all that, Mr. Andrews.

ANDREWS: Ah, I'll talk to her myself. Have some food brought up to her.

CAPTAIN: Yes, sir.

And then the father's object during the ensuing interview is to get his daughter to accept food from him; he even tries to feed her, but she responds as though he is trying to jam something down her throat; and when she deliberately knocks to the floor the tray of food he has had sent up, he slaps her, upon which she runs from the cabin and dives from the yacht to escape him. The angry refusal of food is thus directly established as an angry, intimate refusal of love, of parental protection; the appreciative acceptance of food in the auto camp cabin asserts itself as the acceptance of that intimacy. This relation has also been prepared earlier by the man's having denied food to her on the bus. It occurs just after he has moved out the flirt by claiming to be her husband. Colbert orders a box of chocolates from a vendor on the bus and Gable sends the boy away. He explains this husbandly act by saying that she can't afford chocolates and telling her that from now on she's on a budget; but the implication is clear enough that he is instructing her not merely in what is worth spending but in what is worth eating, say in what is worth consummation. (And in what manner what is worth consuming is worth consuming—recall the lecture on doughnut dunking.)

The next night, in the field, she complains of hunger, and when later the man, after disappearing, returns with a bunch of carrots, she refuses them, saying she's too scared to be hungry. The next morning he offers them again:

ELLIE: What are you eating?

PETER: Carrots.

ELLIE: Raw?

PETER: Uh-huh. Want one?

ELLIE: No! Why didn't you get me something nicer to eat?

PETER: That's right, I forgot. The idea of offering a raw carrot to an Andrews. Say you don't think I'm going around panhandling for you, do you? Better have one of these. Best thing in the world for you, carrots.

ELLIE: I hate the horrid things.

He is exasperated by the irrationality of her refusal of good food, perhaps by the return of her past over her recent show of genuine feeling, and perhaps he would like to jam the good food down her pretty throat.

Ellie's refusal here aligns Peter even more directly with the opening position of her father, and it sets up a repetition of the earlier pair of actions toward food: again Peter denies her something to eat, again for a moral reason, and as a result she soon accepts food that he has provided for her. The man giving them a lift has stopped in front of a lunchroom:

DRIVER: How about a bit to eat?

ELLIE: Oh, that would be love—

PETER: No thanks. We're not hungry.

DRIVER: Oh, I see. Young people in love are never hungry.

PETER: No.

. . .

PETER: What were you going to do? Gold-dig that guy for a meal?

ELLIE: Sure I was. No fooling, I'm hungry.

PETER: If you do, I'll break your neck.

When they get out of the car to walk around and stretch their legs, the driver hurries out of the lunchroom and takes off with their belongings. Peter runs after the car. After a dissolve, Ellie is waiting beside the road and Peter shows up in the car alone. As they drive off Peter asks Ellie to take the things out of the pocket of his coat, which she is holding, to see what they might exchange for gasoline. One thing she finds in the pocket is a carrot, which, after a hesitation, overcoming something, she begins slowly to nibble, hunching down inside herself. Seeing her eating this food of humility, Peter is won to her. He had liked the taste she showed in people (except for the man she got married to, but then as her father had said, she only did that because he told her not to), but he had despised her sense of exemption from the human condition, a sense he calls her money. Eating the carrot is the expression her acceptance of her humanity, of true need—call it the creation of herself as a human being. No doubt he is also won because eating the carrot is an acceptance of him, being an acceptance of food from him. It is also an acceptance of equality with him, since he has been living on that food. (In one discussion of these matters it was pointed out to me that a carrot is a phallic symbol. I confess to feeling sometimes that certain information is after all really better repressed. But in case someone finds himself or herself saddled with this thought of the carrot, I may mention that some decades further down the road of feature-film making, this region of male nurturance, connected with the attempted creation of a woman, and of a perfected society, was explicitly under consideration.)*

* For example in Dušan Makavejev's *Sweet Movie*, an account of which is contained in my essay "On Makavejev On Bergman," *Critical Inquiry*, Winter 1979. I have been variously questioned about my little gag concerning being instructed by the obvious observation about the carrot. Why avoid this blatantly obvious fact, especially since it is consistent with what I do say about the carrot? I suppose my gag registers the continuing ambivalence in me concerning the pervasive problem of the obvious, about when criticism must state the obvious and when it must avert its stating. One does not want to penetrate the sexual censoring too soon or tactlessly, pull the blanket down before it has done its work, by screening, of forming hope—to create the reality of disillusionment before honoring the truth of illusion. The timing of explicitness is a place at which comedy and tragedy and farce and melodrama will find their differences. I am saying that explicitness poses analogous issues for criticism. An interpretation offered at the wrong place, in the wrong spirit, is as useless, or harmful, as a wrong interpretation. My feeling about the carrot is that we have no more use for making its phallic symbolism explicit than Ellie and Peter would have—I mean at the time she accepts it in the car, on the way to their third night together. Surely *we* do not need to be told that their relationship has sexual overtones or undercurrents. To discover this together, and acceptably, is, rather, exactly their problem. And to suppose that this comes down to discovering the carrot's symbolism strikes me as denying the dimensions of significance I have traced in the carrot—its place as a food, uncooked, and as a gift, from a father. If "phallic" is taken to mean, in this context, all of these things, in balance, then I have no objection to saying it out loud. But then are we not duty bound, as interpretors, to consider what is going on when, sitting on the fence while waiting to thumb a ride, the woman complains of something caught between her teeth, called I think a strand of hay, which the man

It is pertinent at this stage of his being won that the food is raw (a point insisted upon by her earlier), which means that he has provided it, out of his masculine capacity, but not prepared it, out of his feminine capacity. Out of *which* masculine capacity—fathering or husbanding? I am reminded here of an observation of Margaret Mead, quoted by Lévi-Strauss in *The Elementary Structures of Kinship:* "An Arapesh boy grows his wife. As a father's claim to his child is not that he has begotten it but rather that he has fed it, so also a man's claim to his wife's attention and devotion is not that he has paid a bride-price for her, or that she is legally his property, but that he has actually contributed the food which has become flesh and bone of her body."**

I adduce this bit of anthropological observation not as a confirmation, in a conventional or professional sense, of what I have been saying. It is just as much the case that what I have provided is confirmation of that observation—or else my experience of the film is inaccurate and improvident.

I quote it rather to help measure a question that is bound sooner or later to make its way to this discussion—namely, whether I am seriously suggesting that Frank Capra is to be understood as intending to draw the distinctions I have invoked between providing and preparing food and the parallels between the woman's two refusals of the offer of food and the pair of denials of food to her, followed by her acceptance of humbler food, and so on. If such a question is asked rhetorically, I might reply that it strikes me as based on a primitive view of who Frank Capra is, or any authentic film director, and a primitive view of what a Hollywood film is, or film generally, and a primitive view of what having an intention is. Or, I might rather say that one would do well to try conceiving of Capra as possessed of as usable a set of intel-

---

removes with his knife? My reference to *Sweet Movie* is not meant simply to suggest that now, four or so decades after the scenes in question, we are in the mature position of being able to treat such things explicitly; it is meant equally to suggest that now we are in the immature position of not being able to treat such things implicitly.

The comic question of the timing of explicitness is most extensively at play in *Bringing Up Baby,* as at once its subject and its manner, the business of the next chapter. It is worth noting, thinking of *Bringing Up Baby,* that the work of Freud's that is most directly pertinent to these comedies is not *Jokes and Their Relation to the Unconscious* but *The Psychopathology of Everyday Life,* thus affirming Freud's insight that parapraxes, like symptoms, like dreams, share the logic, or the psychology, of jokes.

** (Boston: Beacon Press, 1969), p. 487.

IT HAPPENED ONE NIGHT

The woman believed she was walking into the man's dream or vision. So she was, and it woke him up, or brought him to. Why? Is it because she is not the figure of his vision or because she is? Both.

lectual operations as your average primitive mind. Naturally, these impatient replies do not answer questions, raised nonrhetorically, about how to understand what the director of a film is and what his or her intentions are, which, first of all, means to understand what a film is. The primitive mind, the human mind, can mean things because it has the medium of human culture within which to mean them, and mean itself, where things stand together and stand for one another. The genre of remarriage is a small medium of this sort, wherein distinctions can be drawn and, hence, things intended.

The intentions can get reasonably refined. We are about to consider the sequence in which the crisis of the film occurs on the line "Boy, if I could ever find a girl who's hungry for those things...," and I claim that the energy of the emotions we have seen concerning food is concentrated into that idea of hunger. The film can be said to be about what it is people really hunger for, or anyway about the fact that there really is something people hunger for. (Sometimes, it is not denied, this really is literal food, as in the Depression vignette of a mother on the bus fainting from what her crying child informs us is hunger, and Ellie gives the child the bulk of the money she and Peter have between them. One must have a heart of stone to witness Capra's virtuosity in pathos without laughing.) And, one way or another, an early exchange as Peter is preparing their beds of straw in the open field may make itself felt as a summary of what the film is about.

ELLIE: I'm awfully hungry.

PETER: Aw, it's just your imagination.

(Not unworthy of Beckett.) Will it be objected that we can hardly be expected to remember such a transient identification (of hunger with imagination) on just one viewing—particularly, I might add, since the style of such films, of films as such, tends to throw lines of significance away, quite as if transience, hence improvisation, were part of the grain of film. But does this assume that films are on the whole *meant* to be viewed just once? Films such as this one are meant to *work* on just one viewing, but that is something else. (It is not another matter, however, because the issue of the transient speaks to what we are to understand as the popular in art.) However, I wish to be reasonable about the question of intention. I earlier quoted a passage that contains the juncture

How about a bite to eat?

Oh, that would be love—

and to this I am willing to say that, if you don't see in it another announcing of the film's subject (love as the willingness to admit the satisfaction of hunger), I will not insist upon it. I might still go on to ask how far one is prepared to go with Freud's insistence that the life of the mind contains no accidents. If this is modified to say that, in matters native to oneself, one does nothing by accident, then I will simply claim that making Hollywood romances is something native to Frank Capra.

THE WOMAN'S EATING of the carrot closes the sequences of the night in the field and the following day on the road, and prepares for the pair's third night together, their second in an auto camp cabin, which sees the climactic transgression of the blanket-wall-screen-barrier.

We have noted the particular familiarity in which the pair have now become a couple, preparing for bed; and I have called attention to the intimacy of their speaking of never seeing one another again. There ensues an exchange of words that we must hear at length. He is in his bed, smoking, thinking. She is sitting on her bed getting undressed and into her pajamas, that is, his pajamas, as on that distant night before last.

ELLIE: Have you ever been in love Peter?
PETER: Me?
LLIE: Yes, haven't you thought about it at all? Seems to me you could make some girl wonderfully happy.
PETER: Sure I've thought about it. Who hasn't? If I ever met the right sort of a girl, I'd—. Yeah, but where are you going to find her, somebody that's real, somebody that's alive? They don't come that way any more. I've even been sucker enough to make plans. I saw an island in the Pacific once. Never been able to forget it. That's where I'd like to take her. But she'd have to be the sort of girl that'd jump in the surf with me and love it as much as I did. You know, those nights when you and the moon and the water all become one and you feel that you're part of something big and marvelous. Those are the only places to live. Where the stars are so close over your head that you feel you could reach right up and stir them around. Certainly I've been thinking about it. Boy, if I could ever find a girl who's hungry for those things.

And now the crisis. We have cut to close-ups of Ellie two or three times during Peter's speech, and at last we cut to her just coming toward him around the blanket. She pauses, holds on to the blanket's edge, and we reframe to a tighter close-up of her in soft focus, the visual field blurred as if seen through a mist of happiness, or a trance of it. Then she approaches his bedside, falls to her knees, throws her arms around his neck, asks him to take her with him, and declares her love for him. He seems unmoved, paralyzed, and tells her she'd better get back to her bed. She apologizes, hurriedly retreats back to her side of the barrier, throws herself onto her bed, and sobs. The camera has jumped back as she returned to her side, allowing the blanket to be seen from the edge, dividing the screen frame in half, and depicting the full geography of transgression. It seems, in itself, hardly significant, nothing more than an auto camp cabin. Then, a dissolve from her asleep back to him with his cigarette burnt down conventionally states a lapse of some minutes. He calls out, "Hey, Brat. Did you mean that? Would you really go?" It is late for that question, and he seems stupefied. I understand him as trying to awaken from a trance. He gets up and looks over the barrier to discover that she is asleep, and he hurriedly dresses and leaves, we discover, to sell their story to his old editor. He will tell him that he's in a jam and needs a thousand bucks. He will say, mysteriously, that it is to tear down the Walls of Jericho. What is going on?

We understand her well enough. The man's recital of his wish for love is something we have seen penetrate her as she follows it; her body expands with the imagination of what he is envisioning; her head arches back as her eyes close; her state is depicted as openly, as theatrically, as Bernini depicts St. Teresa. She is drawn toward Peter's vision, hence to Peter. (Though someone may rather characterize her the other way around, as being drawn to Peter and hence to his vision. But we have before this had sufficient evidence that she is already drawn to Peter, if this means attracted to him; and what draws her to him has always been something like his vision. The question that remains is what draws her to declare herself. And if this is a matter of asking why she listens to him now, that becomes the question why he speaks now. He speaks of love because she has asked him to. Then why does she ask him now, and why does he answer? It would be an answer to both questions to say that she has accepted the carrot.) The soft focus is a sign of her yielding, that she is tender. The stars in her eyes signal a trance. But are they not the stars present in his words, hence shared, hence objective? It is the man whose behavior is mysterious.

When he becomes aware of her presence, what does he see? Are we to take it that the soft focus is something he sees as well as some way in which he sees it? He sees that some vision qualifies her state, that she is entranced, but it is not evident to him what or how she sees. This seems to be what the closer reframing on her means as she rounds the edge of the barrier: since the first framing of her is presented as from Peter's point of view, it follows that the succeeding reframing is exactly not from that point of view, but rather that what it presents is something still private to her (and us). We have to imagine that Peter surmises something of her mood, since the first and second framings of her are not unrelated (which is what the idea of reframing should convey). It is this that leaves him, that turns him, cold to her—say objective—as she throws herself at him. His focus is going hard again. Here is Capra taking on responsibility for the Hollywood device of soft focus, raising for us the ontological question: If soft focus registers a modification of viewing, how is hard focus different? Is it merely one modification among others, or is it privileged to escape modification altogether? Or should we seek to define it as a modification of viewing but nevertheless a privileged one? Gable's shift of mood from soft to hard provides one interpretation of (what seems to be) ordinary focus, an interpretation of it as cold, or let us say inquisitive. Focus is a necessary condition for viewing film, as is each condition for exposing film. These conditions are no more to be sidestepped in viewing film than the pure forms of sensible intuition (that is, space and time) in *The Critique of Pure Reason*. If we are to find a way to speak of these conditions of viewing film as transcendental, we must equally find a way to speak of them as empirical, for certainly they are only to be discovered empirically, or rather discovered in what I call acts of criticism.

The woman believed she was walking into the man's dream or vision. So she was, and it woke him up, or brought him to. Why? Is it because she is not the figure of his vision or because she is? Both.

That she is the woman of his dreams seems to me specifically announced in his recital of his dream, his expression of it, no more importantly by what he says than by his saying of it to her, in those circumstances. His invocation of "those nights when you and the moon and the water all become one and you feel you're part of something big and marvelous. . . . Where the stars are so close you feel you could reach right up and stir them around" is of something he is wishing for all right, but more directly I take it as something he is recalling, their previous night together, in the open. The transcendentalism of his vision of oneness with the universe is an exact response to the American transcendentalism of Capra's exteriors, a mode of vision inherited from German expressionism, both in the history of Hollywood and in the history of philosophy. (Capra's handling of emotion, or sentiment, what I earlier referred to as his virtuosity of pathos, seems to me rather to bear an Italian stamp—the equivalent, and surely the equal, of Puccini's.) Kant finds that man lives in two worlds, and camera and screen seem an uncanny, unpredictable realization of the human aspiration, or projection, from the one to the other. (I would like film's manifestation of this idea of Kant's to be thought of in connection with my speaking of film as giving to its subjects an inherent self-reflection, a mutual participation of objects and their projections.)*

What happened one night is that the man took the woman to his island. He carries her across a body of water that Capra's camera, in something like soft focus, shows so brilliant with reflected skylight that there seems no horizon, no break between the earth and the heavens, so that you feel you might reach anywhere and stir the stars. And the place he takes her to is as isolated as an island and is home to him, as is shown by his knowing where the carrots are, and they and the moon and the landscape all become one, as movingly a part of something big and marvelous as any expressionist painterly composition by a movie camera can achieve. The way he carries her across the water is emphasized by his lecturing to her to the effect that what they are doing is by no means piggybacking. He is here not treating her, I mean carrying her, as a child, but over his shoulder. If you call this a fireman's carry, then say that it is accordingly a carry of rescue, as of a hardy damsel. And if we shall not refer to it as the style of a caveman, let us at least note that it takes the posture of abduction.

Then we have again to ask why he withdraws from her when she is drawn past the barrier to his side of things; moreover, when her being drawn seems to remove the remaining impediment to the marriage of their minds. In the previous sequence she accepted her relation to common humanity, and in crossing the barrier she accepts the role of Israelite; the initiating sound has come from her side of things. What is the matter? Why, after all, is he surprised by her? Why can he not allow the woman of his dreams to enter his dream? But just that must be the answer. What surprises him is her reality. To acknowledge her as this woman would be to acknowledge that she is "somebody that's real, somebody that's alive," flesh and blood, someone separate from his dream who therefore has, if she is to be in it, to enter it; and this feels to him to be a threat to the dream, and hence a threat to him. To walk in the direction of one's dream is necessarily to risk the dream. We can view his problem as one of having to put together his perception of the woman with his imagination of her. This would be precisely the right

* This idea of "participation" in speaking of the photographic is introduced in the Foreword to the enlarged edition of *The World Viewed*, specifically in connection with Terrence Malick's *Days of Heaven*.

tumbling of the Walls of Jericho. It is a way to frame a solution to the so-called problem of the existence of other minds.* The genre of re-marriage invites us to speak of putting together imagination and per-ception in terms of putting together night and day—say dreams and re-sponsibilities. Each of its instances has its own realization of this project. But the sublimest realization of it in film is Chaplin's *City Lights*.

Surely, it will be said, a simpler explanation for Peter's rejection of Ellie's advances is that legally she has a husband. But, first of all, Peter could have said that; and second, that seems not to bother him a few moments later when Ellie is back on the other side of the barrier. And how would this explanation accord with Peter's leaving to get a thou-sand dollars in order to tear down the Walls of Jericho, and by selling their story? It is a very mysterious nest of actions.

Whatever the actions mean, the fundamental fact about them is that he leaves, he continues his withdrawal from her, he panics. It is that fact that any explanation must explain. I know he has his reasons and his intentions to return and so on (including some reason why he doesn't wake her and tell her his plans or indeed go so far as to take her with him?). Maybe he thinks that after the hard night confessing her love and sobbing she needs a good sleep. As he heads back with his money and his elation (as if he has escaped from something) he says to his car, "Come on baby, come on, we've got to get there before she wakes up." He has wanted her asleep during this escapade. He does not want her to wake up to the fact that he abandoned her when she crossed the barrier to him; he wants to be able to give her a good, daytime reason for his paralysis; so he abandons her again. What he goes on to do must there-fore remain mysterious, as cover stories will. This failure on his part is never fully compensated for. It remains an eye of pain, a source of sus-picion and compromise haunting the happy end of this drama. So, at any rate, I assume in looking further for explanations. (It has been sug-gested to me that I have looked too far already, that it is an explanation of Peter's conduct to say that, like most respectable males in our cul-ture, he feels he cannot live on the woman's money but must have his own; his manhood, his freedom of action, depends upon it. This is roughly Peter's understanding of his behavior, and I do not say that it is

---

* The theme of other minds is dominant in Part Four of *The Claim of Reason*. A version of the closing pages of that part, hence of that book, forms the major part of "Epistemol-ogy and Tragedy: A Reading of *Othello*," in *Daedalus*, Summer 1979.

wrong, merely that it is an explanation this film goes on to ask and to give an explanation for. You can accordingly take the film's question to be: What does this culture understand a provider to be, and what is the cost in it of a genuine desire to be one?)

THERE IS ANOTHER cultural or psychological interpretation of his withdrawal, another impediment that the man may feel the woman has transgressed; one that is not an alternative to her metaphysical transgression in presenting herself as somebody real, somebody alive; but rather one in which the cultural and the metaphysical interpret one another. This is the natural cultural impediment in their having ostensibly based their relationship on the tie between father and daughter, according to which they are not free to take one another as independent sexual objects. But we have all started from familial attachments, and if we are to proceed in satisfaction—call this marriage—exchanges of one mode of love for another are to be negotiated. What is the particular problem of the central pair of *It Happened One Night?*

But having seen that they have a problem, at least that the man has one, why suppose that this is particular, if this means peculiar to them? It may just be peculiar to romance that it studies the problem of the romance of love not as an individual but as a metaphysical problem, projecting characters free of private problems (free of economic struggles, for example), or rather characters whose problem is exactly that of metaphysical privacy. They trace the progress from narcissism and incestuous privacy to objectivity and the acknowledgment of otherness as the path and goal of human happiness; and since this happiness is expressed as marriage, we understand it as simultaneously an individual and a social achievement. Or, rather, we understand it as the final condition for individual and for social happiness, namely the achieving of one's adult self and the creation of the social. (This does not deny that privacy is also a happiness and publicness also a loss, even the publicness of marriage; hence, it does not deny that comedy cannot just be happy.)

If we express the condition of marriage as one in which first a kinship is to be recognized and then an affinity established, we have a possible explanation for a genre of romance taking the form of a narrative of remarriage: a legitimate marriage requires that the pair is free to marry, that there is no impediment between them; but this freedom is announced in these film comedies in the concept of divorce. (In *The Lady Eve*, as remarked, the woman sets as her sole condition for giving the man a divorce that he come and "ask me to be free." I am encouraged in this connection by Bernard Knox's "*The Tempest* and the Ancient Comic Tradition,"* to think through the films of remarriage, especially in their relation to Shakespearean romance, as comedies of freedom.) But then this implies that a prior marriage or affinity is in question. This original affinity may or may not be depicted as a legal one; but it is essential that its originality be seen to be, let us say, a natural one. I have, for example, mentioned that in *The Philadelphia Story* the pair are said twice to have "grown up together"; in *Bringing Up Baby*, as we shall see in detail in the following chapter, they are shown becoming children again (something Howard Hawks pushes to extravagant literality in *Monkey Business*); in *It Happened One Night* the whole escapade of escape may strike one as a set of games, but especially the game of playing house, or playing at being married. But this natural relationship is a kinship from which freedom to marry is precisely to be won. Without the kinship, the eventual marriage would not be warranted; without the separation or divorce, the marriage would not be lawful. The intimacy conditional on narcissism or incestuousness must be ruptured in order that an intimacy of difference or reciprocity supervene.** Marriage is always divorce, always entails rupture from something; and since divorce is never final, marriage is always a trangression. (Hence marriage is the central social image of human change, showing why it is and is not metamorphosis.)

So it is no wonder that Peter is confused by Ellie's appearance to him, and he is not to be blamed for an act of rupture, or abandonment, that he cannot heal. We might understand his leaving her asleep as his intuition that they require, and his going in search of, a divorce; and understand his failure to accomplish this as his discovery that it cannot be accomplished alone. His task reverses, or reinterprets, the story of Sleeping Beauty: the prince wishes for the maiden to stay asleep until

* Reprinted in the Signet edition of *The Tempest*.

** Another observation from *The Elementary Structures of Kinship* bears on this logic. Lévi-Strauss describes a ritual of the Ifugao of the Philippines that prescribes a sham fight to be undertaken between the families of the bridegroom and the bride in order to establish that they are on different sides of the matrimonial fence (pp. 83–84).

he finds his way to the authority to kiss her, but a witch has put a curse on her to awaken, after a hundred years, a moment too soon.

Another witch must have modified the curse so that the faithfulness of the prince attracts the favor of the maiden's father. The magic in the relation between father and daughter is noted in two junctures of *It Happened One Night*, which I take as comic confessions that this film knows its complicity in the tradition of romance. The first is the daughter's being awakened the first morning of her escape by the sound of her father's airplane passing overhead, as if she had been dreaming of him in pursuit. The second is the father's arranging to put a headline in the newspaper, which is then revealed as being in the hands of his daughter; he contrives to get a message to her, not knowing where she is, not by using the personal columns of the paper but by using the medium of the newspaper as such as his personal means of communication with her. This Prospero is playing his own Ariel. (The image of the airplane sets up a further comic moment of contrast with the senex-husband, who insists on entering the wedding ceremony from an "autogyro," an act of vanity, of the merely autoerotic. Yet this dunce *ex machina*, by his "dumb stunt" [as the father calls it], helps the sudden averting of the catastrophe, by displaying what a male catastrophe he is. The spirit of comedy continues to work in mysterious ways.) The role of the fantastically rich father shows the limitation of the (modern) father's powers. He can no longer *give* the bride, he can merely use his personal standing with her to persuade her to give herself to a better man. When he says he can buy off her legal husband with a pot of gold, he is admitting that a true husband cannot (any longer) be bought, perhaps because the honor of a modern husband lies not in the greatness of his price but in his having no price, as if we are no longer in a position to tell the difference between a gift and a bribe. Even if reduced in this way, however, this father participates in his ancient role of magus in ousting the senex, in lending what authority he has to his daughter's flouting of convention. His limitation also emphasizes that she is, as France says of Cordelia, her own dowry. Otherwise, there is no romance.

To be your own dowry turns out to mean to give yourself. And the woman gives herself to the man who genuinely wins her. The woman's giving herself looms so large as a feature of modern (European) marriage that I cannot forbear giving a final citation from Lévi-Strauss that speculates on its origins. It has a foundation in

> the *swayamvara* marriage, to which a whole section of the Mahabharata is devoted. It consists, for a person occupying a high social rank, in the privilege of giving his daughter in marriage to a man of any status, who has performed some extraordinary feat, or better still, has been chosen by the girl herself . . .[S]wayamvara, the marriage of chance, merit, or choice [rather than of gift by kindred], can really only have meaning if it gives a girl from a superior class to a man from an inferior class, guaranteeing at least symbolically that the distance between the statuses has not irremediably compromised the solidarity of the group, and that the cycle of marriage prestations will not be interrupted. This is why the lower classes have a major interest in the *swayamvara*, because for them it represents a pledge of confidence.*

This natural explanation for the popularity of these comedies also provides terms for an account of two respects in which *It Happened One Night* differs from each of the other main instances of the genre of remarriage. First, the man is emphatically from a lower social rank than the woman. Second, there is nothing that corresponds to the feature of "the green world." Instead, we have here a picaresque form, the narrative is on the road, and instead of memory and reciprocation, *adventurousness* is given the scope it requires in order to win out. (What "instead" comes to here is sketched out, using this example, in the Introduction.)

THAT THE BARRIER works like a movie screen means that our position as audience is to be read in terms of the man and woman's positions with each other, and especially in terms of the man's, for it is with him that we first watch the screen take on the characteristics of a movie screen, and his problem of putting together a real woman with a projected image of her seems a way of describing our business as viewers. I urged in Chapter 1 that the insistence on the identity of the female star in the films of our genre presents analogous issues for the characters and for their audience, and is a way of insisting on the cinematic status of the film we are viewing. In *It Happened One Night* a subordinate, or personal,

---

* *The Elementary Structures of Kinship*, pp. 472, 476.

assertion of the reality of Claudette Colbert is her stopping a car by showing her leg, one of the two most famous events in the film (the other is the companion piece of Gable removing his shirt) apart from the blanket-wall itself. When these very stockings she straightens, through which she affords this revelation (the clothes she is wearing are all she has, the rest were stolen with her suitcase), were laid by her across the top of the Walls of Jericho soon after it was first constructed, the man was stirred by the imagination of them. But now, seeing the stockings on the live woman, he is apparently left cold. Of course, the timing and the placing do not exactly set the scene of desire, even, I would imagine, for a fetishist. And yet, I think there is more to his disapproval of her gesture than an expression of anger at the spectacle she has made of herself, and more than his knowledge that her resourcefulness has in this instance outstripped his. In the succeeding shot, as they are shown being driven along, his gloominess as well as her lightness of spirit proceed as well from her having insisted, for his benefit, upon her physical identity. (I know of no one who fails to find Gable an extraordinarily charming and powerful presence in this film, even to find his intelligence and wit something of a revelation, given the way he is seen in his more explicitly adventure films. But a number of people, mostly it seems to me, women, cannot warm up to Colbert. She is, I believe, sometimes found brittle, or shallow, and in any case too obscure in her emotions to warrant and satisfyingly return the attentions paid her by Gable. I have felt this, but my settled view of her in this film is that her performance, though something less than Gable's, is yet a fine match for it. It may be that Colbert lends herself to a kind of abstract view, so that one takes oneself to know what she is like without really looking at her concretely and in detail. I suggest that one make an effort to force oneself away from Gable for a moment and pay special attention to her throughout the sequence of the marriage charade and to the way she delivers her lines before and after she exhibits her leg ["I'll stop a car and I won't use my thumb," "It proves that the limb is mightier than the thumb"]. But I do not wish to insist upon this. I merely say that if you can't respond to Colbert's individuality in something like the way one responds to Gable's, this film is going to have trouble bearing the weight I place on it.)

The Lady Eve employs the most extreme device in its demonstration that acknowledging the identity of the woman requires putting two versions of her together, one in which the woman apparently plays her own twin. *It Happened One Night* employs the most economical device in its demonstration that this putting together is a spiritual task, a task of education. (This is a way of summarizing different emphases that show up between one film that declares the nature of a film screen by holding up a mirror and another that declares it by hanging up a blanket.) Both films further specify their own presence as films by suggesting that one of the pair is a surrogate for the film's director (where this role is not particularly distinguished from the film's author, if this means its screen writer). In *The Lady Eve* it is the woman who directs the action (as it is in *Bringing Up Baby*); the man is her audience, gulled and entranced as a film audience is apt to be. In *It Happened One Night* it is the man who directs, and the woman is not so much his audience as his star. The main bits of evidence for Peter as director are, of course, the events that allow us to interpret his creation of the barrier as the setting up of a screen. And we have remarked on his invention of their stagings of marriage. (I will not dwell on it, but Peter's treatment of Ellie in getting her ready to act the abused wife, the extraordinary gestures in which he musses her hair and unbuttons her blouse and rearranges her skirt at her knees, is at once a declaration of his parental powers but also of something else between them as out of the ordinary as anything in film I know of—I am at the moment calling this his direction of her.) Recall as well that in telling her of his reasons for their sharing one cabin and his insistence on following her around for her story, he has said: "And if you get tough, I'll just have to turn you over to your old man right now. Savvy? Now that's my whole plot in a nutshell. A simple story for simple people." The next morning when she says he thinks her running away is silly, he denies it, saying that it is too good a story, thus distinguishing her story from his plot, which is to force her to yield up her story "exclusive." Ellie's reply is to be considered. "You think I'm a fool and a spoiled brat. Perhaps I am, but I don't see how I can be. People who are spoiled are accustomed to having their own way. I never have. On the contrary, I've always been told what to do and how to do it and where and with whom." This is a reasonably literal description of what it is like to be a film star, and no less an accurate description of the way Peter treats her.

We have available a reading of the allegorical confusion Peter suffers in leaving Ellie while he goes to sell their story. The telling of the story is to have the effect of authorizing their marriage, of divorcing them from their pasts, she from her father and her legal husband, he from his private fantasy; of making something public. He calls it "the biggest scoop of the year." But what is the scoop, what is *their* story? He tells the editor he needs the thousand to tear down the Walls of Jericho and that her marriage is going to be annulled so that she can marry someone else, namely him. But that is *not* news, because his abandoning her means precisely that he does not know, or have it on good authority, that they can marry. He is behaving as though announcing the event in the newspaper will not only make it public but make it happen. And maybe this could work, for some story other than his own. When the editor acquires conviction in the story, from the authority of the story itself, in the face of clear evidence against it, what he says to Peter in consolation is that, with a great yarn, something always comes along and messes up the finish. This would presumably be reality. It is not Peter who has messed up the finish of the yarn; on the contrary, the only yarn is the one he has written for the editor, and the editor accepts the finish. He accepts what we have accepted. Their story has already happened; it cannot be *made* public if *this* is not public. What has made it public is a film, and in that sense a yarn. A newspaper story, coming after the fact, has no further fact to come after here. That the yarn that has happened has no assured finish in the future happiness of the pair, and is in that sense messed up, is part of the logic of the human work of construction, of art, of its transience and its permanence, as John Keats and Wallace Stevens best say; and part of the logic of the human emotion of love, what Freud calls its biphasic character; not something that this man is singularly to be blamed for.

It is notable that Peter never does transgress the barrier. When Ellie does not respond to his belated question about whether she really means what she has so passionately declared, he looks over its top, and, when he is dressed to leave, blows a kiss over it. As a man he is merely playing out the string of his confusion. Seen as a surrogate for the film's director, he is playing out the condition of film and its subjects, that their maker has to kiss them goodbye, that he or she is outside, that when the play is done his work is absolutely over, unlike the case of theater. He has become the work's audience, the viewer of creation, its first audience, but with no greater power over it therefore than any later one. He puts himself in the hands of higher powers, not unlike the duty of a romantic hero. He has accomplished his remarkable feat. His reward must be *conferred*.

I said earlier that in working like a movie screen the barrier represents both an outer and inner censoring, and more recently that the man's problem in connecting the woman's body and soul, that is in putting together his perception and his imagination, his and her day and night (so that his capacity for imagination becomes his ability to imagine *her*), is a framing of the problem of other minds. Putting these notions together in turn, I would read the instruction of the barrier along these lines. What it censors is the man's knowledge of the existence of the human being "on the other side." The picture is that the existence of others is something of which we are unconscious, a piece of knowledge we repress, about which we draw a blank. This does violence to others, it separates their bodies from their souls, makes monsters of them; and presumably we do it because we feel that others are doing this violence to us. The release from this circle of vengeance is something I call acknowledgment. The form the man attempts to give acknowledgment is to tell their story. The film can be said to describe the failure of this attempt as a last attempt to substitute knowledge for acknowledgment, privacy for community, to transcend the barrier without transgressing it. Only of an infinite being is the world created with the word. As finite, you cannot achieve reciprocity with the one in view by telling your story to the whole rest of the world. You have to act in order to make things happen, night and day; and to act from within the world, within your connection with others, forgoing the wish for a place outside from which to view and to direct your fate. These are at best merely further fates. There is no place to go in order to acquire the authority of connection. The little community of love is not based on the appeal of the law nor on the approach of feeling. It is an emblem of the promise that human society contains room for both, that the game is worth the candle. You cannot wait for the perfected community to be presented. And yet, in matters of the heart, to make things happen, you must let them happen.

3

LEOPARDS
IN
CONNECTICUT

Bringing Up Baby

*The principals' actions consist of, or have the quality of,
a series of games; the female of the pair likes the games
whereas the male plays unwillingly; their behavior is
mysterious to everyone around them.*

SPEAKING of the relationship of the principal man and woman of our comedies as one, as revealed in *It Happened One Night*, in which what they do together is less important than the fact that they do whatever it is together, I said that Howard Hawks's *Bringing Up Baby* (1938) is the member of the genre that presents the purest example of this quality of the relationship. I called this quality the achievement of purposiveness without purpose (or say directedness without direction). In thus invoking Kant's characterization of the aesthetic experience I am thinking of his idea as providing an access to the connection of the aesthetic experience with the play of childhood, a connection to whose existence many aestheticians have testified. This is not to recommend that we take an aesthetic attitude toward our moral lives; this would not overcome our distance from childhood and its intimacies, but merely cover one distance with a further one. The idea is rather to measure our capacity for perception by the condition of childhood, as for example Wordsworth does, or Freud. I am reminded here of the poignant concluding words of Freud's *Wit and Its Relation to the Unconscious*: "the mood of childhood, when we were ignorant of the comic, when we were incapable of jokes and when we had no need of humor to make us feel happy in our life."

The fact of remarriage between the central pair is even less directly present in *Bringing Up Baby* than in *It Happened One Night*. In the essay to follow I justify its inclusion in the genre of remarriage by emphasizing the pair's efforts to extricate their lives from one another, in which the attempt at flight is forever transforming itself into (hence revealing itself as) a process of pursuit. I should like to add that this transformation can be said to provide the structure of the tale *Gradiva: A Pompeiian*

Fancy, by Wilhelm Jensen, the work of fiction to which Freud allotted his most extended consecutive interpretation. It is pertinent for us that Freud's interest in this romance would have been elicited by its being a tale of the therapy of love. It is the woman who provides this therapy by virtue of her knowledge, whatever the man may think, that she is the object of his (repressed) desire, and her ability to bring him back to this knowledge by virtue of her willingness for the time to live out his delusions (call this sharing his fantasies). The therapy of love provided by the woman making an initial marriage possible (as though women can bear up, where men buckle, under the injunction not to look back, as if either they trust the past or else they can look at it without distorting it—as if they do not succumb to skepticism about love) is a condition that will perhaps not fully manifest itself in these chapters until the final one, on *The Awful Truth*. But it is well to leave the idea as a current underlying the repeated emphases on the saving education provided by the man, which makes the remarriage possible.

To include the principal pair in *Bringing Up Baby* among the pairs in remarriage comedies is, put otherwise, to imply that their conversation is such, their capacities for recognition of one another are such, that what they are is revealed by imagining them as candidates for the trials of remarriage—as though we are here in the earliest phases, say the prehistoric phases, of the myth I began sketching in the Introduction, something I claimed represented an inheritance in which we must conceive the members of a genre to share. I conclude these transitional short subjects by remarking that it sweetens my sense of relevance that the title *Bringing Up Baby*, while suggesting something about the etiquette of conversation, is directly that of an education manual, one of those cute ones, written for the millions who find it reassuring to be told that babies are not scary and mysterious, and that a brand new baby and a brand new parent will naturally educate one another, with no difficult decisions ever having to be made. (Or maybe this mode of discourse is now confined to modern sex manuals.) But it is time for the movie.

IT'S THE ONE that opens in a museum of natural history where an absent-minded professor (Cary Grant) is trying to finish his reconstruction of the skeleton of a brontosaurus. Standing as it were before the curtain, he finds out, or is reminded of, five or six things: that the expedition has just found the crucial bone, the intercostal clavicle, to complete the skeleton; that he is getting married tomorrow to his assistant, Miss Swallow; that after their wedding there will be no honeymoon; that the reconstructed skeleton will be their child; that he has an appointment to play golf with a Mr. Peabody and discuss a donation of a million dollars to the museum; and that he is to remember who and what he is. Call this Prologue the first sequence. There is a natural breakdown into ten further sequences. (2) On a golf course, the professor is drawn from his game and conversation with Mr. Peabody by a young woman (Katharine Hepburn) who first plays his golf ball and then dents and rends his car unparking it, amused at his claim that she has made a mistake and that the car, too, belongs to him, and then drives off with it while he is hanging onto its side as perhaps the bull did with Europa. The sequence ends with his yelling back for the third or fourth time: "I'll be with you in a minute, Mr. Peabody." (3) At night, in the restaurant of some Ritz Hotel, Grant slips on an olive dropped by Hepburn and sits on his hat on the floor. Their argument is resumed concerning who is following whom. After further parapraxes, each rips open part of the other's clothing: she splits the tails of his swallow-tail coat up the back and he rips off the back of the skirt of her evening dress. He walks out behind her, guiding her, to cover what he's done (not, however, what he's doing). As he does so, Mr. Peabody appears again, with whom he again had an appointment, and again he says, "I'll be with you in a minute, Mr. Peabody." (4) In her apartment Hepburn sews Grant's tails, after which they set out to find Mr. Peabody, whom she knows and whom she throws stones at after giving Grant his second drive around. They are on a first name basis by now. David tells Susan that he's getting married tomorrow. (5) The prehistoric bone is delivered to Grant's apartment and he rushes to hers, the bone in a box under his arm, to save her from a leopard, who turns out to be Baby, a tame present from her brother. Susan and Baby arrange that the leopard is not to be Susan's problem alone. (6) Driving Baby to Susan's house in Connecticut, they hit a truck of fowls, buy thirty pounds of raw meat, and Susan steals, this time quite consciously, another car. (7) At the house, Susan does not rip David's clothes off but steals them while he is showering. So David puts on Susan's negligee, and later is discovered in bits of her brother's riding habit, which is appropriate since they soon have to hunt for something rare and precious, the bone which the dog George has taken from the box on the bed. David says to Susan's Aunt (May Robson) that he went gay all of a sudden. He learns that the aunt is the potential donor of the million and that Susan expects to inherit it. He asks Susan most earnestly not to tell her aunt who he is. Susan tells her that he's had a nervous breakdown and that his name is Bone, and that is what the Aunt tells her friend the Major (Charles Ruggles) who appears for dinner. (8) The four are at dinner during which David stalks George. The Major gives the mating cry of the leopard, which is answered. He asks, "Are there leopards in Connecticut?" (9) Baby escapes, George disappears, and David and Susan spend most of the night exploring the woods. Susan enjoys it. They are captured, she by a recurring psychiatrist, he by a recurring sheriff. (10) They are behind bars; eventually most of the household is, from trying to identify them. Susan talks her way out of her cell, then out a window, to get the proof that there really is a leopard in Connecticut. She returns dragging a circus leopard behind her, whom we know to be a killer. David does what he once ran to her apartment to do— saves her from a wild beast. (11) In the Epilogue, back in daylight at the museum, Susan shows up, having recovered the bone and inherited the money. Running up high ladders, they talk across the back of the brontosaurus; he says he thinks he loves her. He rescues her again as she has jumped from her swaying ladder onto the brontosaurus, pulling her by one arm up to his ledge as the skeleton collapses under her weight. They embrace.

AT SOME POINT it becomes obvious that the surface of the dialogue and action of *Bringing Up Baby,* their mode of construction, is a species of more or less blatant and continuous double entendre. The formal signal of its presence in the dialogue is the habitual *repetition* of lines or words, sometimes upon the puzzlement of the character to whom the line is addressed, as though he or she cannot have heard it correctly, sometimes as a kind of verbal tic, as though a character has not heard, or meant, his own words properly. I qualify this presence of doubleness thus heavily (calling it a "species" and claiming that it is "more or less blatant") for two reasons.

(1) While an explicit discussion, anyway an open recognition, of the

film's obsessive sexual references is indispensable to saying what I find the film to be about, I am persistently reluctant to make it very explicit. Apart from more or less interesting risks of embarrassment (for example, of seeming too perverse or being too obvious), there are causes for this reluctance having to do with what I understand the point of this sexual glaze to be. It is part of the force of this work that we shall not know how far to press its references.

At some juncture the concept and the fact of the contended bone will of course threaten to take over. (Its mythical name, the intercostal clavicle, suggests that it belongs to creatures whose heads are beneath their shoulders, or anyway whose shoulders are beneath at least some of their ribs.) This threat will occur well before the long recitative and duet on the subject (beginning with Grant's thunderous discovery of the empty box and the lines "Where's my intercostal clavicle?" "Your *what?*" "My intercostal clavicle. My bone. It's rare; it's precious," and continuing with Hepburn's appeal to the dog: "George. George. David and Susan need that bone. It's David's bone"; hence well before the quartet on the words "Mr. Bone," a title that both claims Grant as the very personification of the subject at issue (as someone may be called Mr. Boston or Mr. Structuralism) and suggests, pertinently, that he is an end man in a minstrel show.

By the close of the sequence in the restaurant, the concept and the fact of the behind will be unignorable. Neither the bone nor the behind will give us pause, on a first viewing, in Grant's opening line, the second line of the film: gazing fixedly down at a bone in his hand he says: "I think this one belongs in the tail." His assistant, Miss Swallow, corrects or reminds him: "You tried that yesterday." That we are not given pause on a first viewing means both that this film is not made for just one viewing and also that this early line works well enough if it underscores the plain fact that this man is quite lost in thought, and prepares us for amazement when we discover what it is he is lost in thinking about, and for discovering that his preoccupation is the basis of the events to come. This is not asking too much. The broad attitude of this comedy is struck at once, at Miss Swallow's opening line, "Sh-h-h. Professor Huxley is thinking," as the camera rises to discover Cary Grant in the pose of Rodin's *The Thinker*, a statue already the subject of burlesque and caricature. (The rightness in its being Cary Grant who takes this pose is a special discovery of Howard Hawks's about Grant's filmic presence, his photogenesis, what it is the camera makes of him. What Grant is thinking, and that what he is doing is thinking, is as much the subject of *His Girl Friday* as it is of the time he reverts to playing professor, in *Monkey Business*.)

Then are we to pause over the lines started by Grant to Hepburn when they discover that Baby has escaped?: "Don't lose your head." "My *what?*" "Your head." "I've got my head; I've lost my leopard." And how much are we to do with Hepburn's line, genuinely alarmed, to Grant as he is trying to cover her from behind in the restaurant? "Hey. Fixation or no fixation . . . Will you stop doing that with your hat?" (What does she think he is doing and what does she think he should be doing it with?) And we are to gasp as Hepburn, in the last scene before the Epilogue, in jail, drops what she calls her "society moniker" and puts on a society woman's version—or a thirties movie version—of a gun moll, drawling out, in close-up: "Lemme outta here and I'll open my puss and shoot the works." I say we do not know how far to press such references, and this is meant to characterize a certain anxiety in our comprehension throughout, an anxiety that our frequent if discontinuous titters may at any moment be found inappropriate. If it is undeniable that we are invited by these events to read them as sexual allegory, it is equally undeniable that what Hepburn says, as she opens the box and looks inside, is true: "It's just an old bone." Clearly George agrees with her. The play between the literal and the allegorical determines the course of this narrative, and provides us with contradicting directions in our experience of it.

(2) The threat of inappropriateness goes with a slightly different cause of my reluctance to be explicit, namely that the characters are themselves wholly unconscious of the doubleness in their meaning. This is a familiar source of comic effect. But so is its opposite. In particular, the effect here contrasts specifically with Shakespearean exchanges in double entendre, where the characters are fully conscious of the other side of their words. The similarity between our characters and comparable ones in Shakespeare is that the women in his plays are typically virgins and the men typically clowns. They are, that is to say, figures who are not yet (or by nature not) incorporated into the normal social world of law and appropriateness and marriage and of consonant limitations in what we call maturity and adulthood.

The critical problem in approaching these characters, or the problem

If it is undeniable that we are invited by these events to read them as sexual allegory, it is equally undeniable that what the woman will say is true: "It's just an old bone."

in describing them, can then be put this way: If we do not note the other side of their words and actions, then we shall never understand them, we shall not know why the one is in a trance and the other in madcap. But if we do note the other side of their words and actions, we shall lose our experience of them as individuals, we shall not see their exercises of consciousness. We have neither to know them nor to fail to know them, neither to objectivize nor to subjectivize them. It is a way of defining the epistemological problem of other minds.

Let us note some further features of the world of this film that there should be no reluctance or difficulty in making explicit. Not surprisingly, given that the film is some kind of comedy, it ends with a marriage, anyway with a promise of marriage, a young pair having overcome certain obstacles to this conclusion. Apart from these central characters, we have a cast of humors—an exasperated aunt; a pedant (in the guise, not uncommon in Hollywood films, of a psychiatrist); a sex-less zany who talks big game hunting; an omni-incompetent sheriff; a drunken retainer—none of whom can act beyond their humorous repetitions. The exposition of the drama takes place, roughly, in the town, and is both complicated and settled in a shift to the countryside. It carefully alternates between day and night and climaxes around about midnight.

Are we beginning to assemble features whose combination, could we find their laws, would constitute a dramatic genre? And should such a genre be called "a Hollywood comedy"? This seems unpromising. Not all the considerable comedies made in Hollywood will contain even the features so far mentioned; and the label hardly captures our intuition that the mood, to go no further, of this film is quite special. Yet Northrop Frye, in an early statement of his work on comedy, allows himself to say: "The average movie of today [he is writing in 1948] is a rigidly conventionalized New Comedy proceeding toward an act which, like death in Greek tragedy, takes place offstage, and is symbolized by the final embrace." This is a nice example of academic humor, and strikes a conventional note of complacency toward movies in general. But is it true?

I cannot speak of the "average movie" of 1948 or of any other time, but of the Hollywood comedies I remember and at the moment care most about, it is true of almost none of them that they conclude with an embrace, if that means they conclude with a shot of the principal characters kissing. It is, in particular, not the way the other members of our genre conclude.

So let us not speak hastily and loosely of final embraces and happy endings. There are few festivals here. The concluding moments I have cited are as carefully prepared and dramatically conclusive (if, or because, fictionally inconclusive) as the closing of an aphorism, and it may be essential to a certain genre of film comedy that this should be so.

Bringing Up Baby, it happens, does conclude with an embrace, anyway with some kind of clinch. It is notably awkward; one cannot determine whether the pair's lips are touching. And it takes place on the platform of a work scaffold, where the film began, and in the aftermath of a collapsing reconstructed skeleton of a brontosaurus. What act does *all* of this symbolize? The collapsing of the skeleton poses the obvious discomfort in this conclusion, or shadow on its happiness. One is likely to ask whether it is necessary, or positively to assert that it is not. Is it meant to register the perimeter of human happiness, or the happenstance of it—like the breaking of the glass at the end of a Jewish wedding? Both surely comment upon the demise of virginity, but in this film it is the woman who directly causes it. Perhaps, then, our question should be, not whether it is necessary, but how it is that this man, afterwards, can still want to embrace. Are we to imagine that his admission of love requires that he no longer care about his work? Or can we see that he finally feels equal to its disruption and capable of picking up the pieces?

It should help us to recognize that the pose of the final clinch—something that to me accounts for its specific awkwardness—is a reenactment of a second popular statue of Rodin's, *The Kiss*; a concluding *tableau vivant* to match the opening one.—So what? Are we accordingly to conclude that the opening man of stone after all retreats into stone? But surely the intervening events have produced some human progress, or some progress toward the human? At least he now has company. The isolation of the scaffold has emphatically become the isolation of a pedestal. It looms so large and shadowy in the final shot as to mock the tiny figures mounted on it. Surely they will make it down to earth?—How did they get up there? It started as Hepburn entered the museum holding the recovered bone, upon which Grant instinctively ran up the scaffold—perhaps *in order* to be followed up. In any case he does at least acknowledge, over the skeleton, that he ran because he is afraid of her, which prepares his declaration of love. So he, or his prehistoric instinct, was as much the cause of the collapse of science as she was; as much the cause of its collapse as of its construction.

The issue of who is following whom presides over their relationship from its inception. At the end of the first scene, on the golf course, he responds to her accusation by denying that he is following her, and in the conventional sense he is not; but it cannot be denied that literally he is. Whereupon she gallops off with him. (She does this again later, and again in a stolen chariot, after their stop in Connecticut to buy food for Baby.) At the close of the restaurant sequence, their walk-off—the man leading the woman yet following her pace, as in some dream tango, dog fashion—identifies the issue of who is following whom with the matter of who is behind whom, which remains thematic in subsequent scenes.

Notably, as the two are hunting through the night woods for Baby, Grant with a rope and a croquet mallet, Hepburn with a butterfly net, he turns around to discover her on all fours (she is trying to avoid his wake of branches swinging in her face) and he says, "This is no time to be playing squat tag"; she replies that she is not playing and, upon asking whether she shouldn't go first, is told, "Oh no. You might get hurt." The question of who belongs where reaches its climax inside the jailhouse in the last scene before the Epilogue. We will get to that.

I have suggested that the work of the romance of remarriage is designed to avoid the distinction between Old and New Comedy and that this means to me that it poses a structure in which we are permanently in doubt who the hero is, that is, whether it is the male or the female, who is the active partner, which of them is in quest, who is following whom. A working title for this structure might be "the comedy of equality," evoking laughter equally at the idea that men and women are different *and* at the idea that they are not. The most explicit conclusion of this theme among the films I can recognize as of this genre is arrived at in *Adam's Rib*. Once more we are in the expensive Connecticut countryside; once more the pair is alone. And we are given what sounds like a twice-told, worn-out joke. Tracy says: Vive la différence. Hepburn asks: What does that mean? Tracy replies: It means, Hooray for that little difference. Then they climb behind the curtains of a fourposter bed and the film concludes. If what I have claimed about the conclusions of such films is correct, then a film so resourceful and convincing as *Adam's Rib* cannot vanish on the sounding of a stale joke. And it does not. It vanishes with a joke on this joke. It is not conceivable that this woman—to whom Tracy had cracked earlier, when she was turning on a superior note, "Oh. Giving me the old Bryn Mawr business, eh?"—it is not conceivable that this woman does not know what the French words mean. She is asking solemnly, what difference is meant by that little difference. So it is upon the repetition of a question, not upon the provision of an answer, that they climb together out of sight into bed, with, surrealistically, their hats on. (How their hats get put on makes a nice story. Her putting hers on is a reacceptance of an important and intimate present from him. His putting his on acknowledges that hers is on. He puts his on without thinking, as another man would take his off in the presence of a lady. This pair is inventing gallantry between one another.)

The equality of laughter at the idea of difference is enough to ensure that, unlike the case of classical comedies, there can in general be no new social reconciliation at these conclusions, for society does not regard the difference between men and women as the topic of a metaphysical argument; it takes itself to know what the difference means. So the principal pair in this structure will normally draw the conclusion on their own, isolated within society, not backed by it. The comedy of equality is a comedy of privacy, evoking equal laughter at the fact that they are, and are not, alone. In particular, the older generation will not be present. Where this rule seems to be infringed, say in *The Philadelphia Story*, the moment is radically undercut; we are ripped from our supposed presence at this wedding festival by being shown that we are looking at a gossip shot—one way of looking at a movie—giving us the sort of inside knowledge that merely underlines our position as members of an outside public. Contrariwise, the pull of the private conclusion can mislead a director into supposing that his picture has earned it. I am thinking of Cukor's *Holiday*, which he concludes with a kiss. This conclusion feels wrong, feels like violation, every way you look at it— from Grant's point of view, from Hepburn's, but especially from the point of view of their older friends, a couple who in this case, themselves being shown out of sympathy with the conventional world, have provided an alternative social world for this young pair and who therefore deserve to be present, whose presence therefore feels required. I mention this in passing partly to enlist another item of evidence for investigating the idea of the final embrace, but also to suggest that the wrongness of this conclusion cannot be accounted for by appealing to a lack in the psychological development of the characters (their development is complete), nor excused by appealing to a general movie convention of the final embrace, first of all because there is no such general convention, and second, and more important, because the wrongness in question consists in breaking the structure of this narrative.

Is there present a definite structure of the kind I have named the comedy of equality? And if there is, what has it to do with the thematic or systematic allegory in *Bringing Up Baby*? How does it help us to understand who or what Baby is, and where a Baby belongs, and where a Baby comes from?

I might bypass my intuition of a definite structure in force here and directly seek an interpretation of the mode of sexuality in play, in par-

LEOPARDS IN CONNECTICUT

In their final game (playing tamer and rescuer), the woman stands behind the man, and, after their victory, he turns to face her, tries to say something, and then loses consciousness, collapsing full-length into her arms for their initial embrace.

ticular, of the ambivalence or instability in it: the situation between this pair cannot remain as it is. Here I would wish to put together the following facts: first, the texture of certain speeches and actions that I have noted as a play between their literal and their allegorical potentialities; second, the sense that the principals' actions consist of, or have the quality of, a series of games (from actual golfing, to rock-throwing at the windows of the rich, to various species of follow-the-leader and hide-and-seek, to playing dress-up and playing house, to finding the hidden object, all framed by pinning the tail on the brontosaurus); third, the fact that the female of the pair likes the games whereas the male plays unwillingly and is continuously humiliated by their progress; fourth, the mystery of their behavior to everyone around them (to Mr. Peabody, from before whose eyes Grant is continually disappearing; to the man's fiancée and to the woman's aunt and to the aunt's major and her cook's husband; to the psychiatrist and the sheriff; and even to the butcher from whom Grant orders thirty pounds of meat for Baby to eat raw).

Such facts add up to a representation of a particular childhood world, to that stage of childhood preceding puberty, the period Freud calls latency, in which childish sexual curiosity has been repressed until the onset of new physiological demands, or instincts, reawakens it. In this film we are attempting to cross the limit of this stage, one whose successful and healthy negotiation demands a satisfaction of this reawakened curiosity, a stage at which the fate of one's intelligence, or ability and freedom to think, will be settled. This stage is confirmed by the air of innocence and secrecy between the two; by the obviousness of the sexuality implied, or rather by the puzzles of sexuality seeming to concern merely its most basic mechanics; and by the perception we are given of the humorous collection of figures surrounding them, a perception of these figures as, one might simply say, grown-ups—not exactly mysterious, yet foreign, asexual, grotesque in their unvarying routines, the source primarily of unearned money and of unmerited prohibitions.

This representation of this period implies two obstacles in the way of this pair's achieving some satisfactory conclusion in relation to one another and to the world, a conclusion both refer to as "marriage." Or, two questions stand in the way of the man's awakening from his entrancement ("I can't seem to move") and of the woman's doffing her madcap ("I just did whatever came into my head"). One question is: If adulthood is the price of sexual happiness, is the price fair? If the grown-ups we see around us represent the future in store for us, why should we ever leave childhood? A second question is: If virginity at some point becomes humiliating and laughable, then why must departing from it be humiliating and laughable? Why are the vaunted pleasures of sexuality so ludicrous and threatening? In the middle of their chase through the woods, they come upon Baby and George growling and rolling in one another's arms on a clear, moonlit patch of ground. Thus seeing themselves, the female is relieved ("Oh look. They like one another"—but she had earlier said that she doesn't know whether, having been told that Baby likes dogs, that means that he is fond of them or eats them); the male is not happy ("In another minute my intercostal clavicle will be gone forever"). I think it would be reasonable, along such lines, to regard the cause of this comedy as the need, and the achievement, of laughter at the physical requirements of wedded love, or, at the romance of marriage; laughter at the realization that after more than two millennia of masterpieces on the subject, we still are not clear why, or to what extent, marriage is thought to justify sexual satisfaction. (That such comedies are no longer made perhaps means that we have given up on this problem, or publicized it to death.) Accordingly, we should regard the midsummer's eve in the Connecticut forest not as the preparation for a wedding ceremony but as an allegory of the wedding night, or a dream of that night. Grant, sensing his entrancement, at one point almost declares himself asleep: "What I've been doing today I could have done with my eyes shut." (At the beginning of the end of the Ritz sequence, he had said: "Let's play a game. I'll close my eyes and count to ten and you go away.") And just before they discover Baby's escape and leave for the woods, he behaves as if he is walking in his sleep, rising stiffly from the dinner table and following George out of the house, his soup spoon still in his hand, stopped in midair on the way to his mouth.

But while I find such considerations pertinent, they seem to me to leave out too much, in particular they do not account for the beginning and the ending of this narrative, for why just this couple finds just these obstacles on their road to marriage. More particularly, they do not account for the overall drive of the plot, which appears to be a story not of a man seeking marriage but of a man seeking extrication, divorce. One might say that according to this plot he is seeking extrication from Hepburn in order to meet his engagement with Miss Swallow. But that hardly matches our experience of these events, which could just as well be described, attending to the introductory sequence, as his attempt to extricate himself from Miss Swallow, who promises him, or threatens him with, a marriage that, as she puts it, must "entail [that word again] no domestic entanglements *of any kind.*" Upon which promise, or threat, he leaves to seek his fortune.

The film, in short, poses a question concerning the validation of marriage, the reality of its bonding, as that question is posed in the genre of remarriage comedy. Its answer participates in, or contributes its particular temperament to, the answer of that structure—that the validity of marriage takes a willingness for repetition, the willingness for remarriage. The task of the conclusion is to get the pair back into a particular moment of their past lives together. No new vow is required, merely the picking up of an action which has been, as it were, interrupted; not starting over, but starting again, finding and picking up the thread. Put a bit more metaphysically: only those can genuinely marry who are already married. It is as though you know you are married when you come to see that you cannot divorce, that is, when you find that your lives simply will not disentangle. If your love is lucky, this knowledge will be greeted with laughter.

Bringing Up Baby shares, or exaggerates, two of the features of this structure. First, it plots love-making in the form of aborted leave-taking. It adds to this, more particularly, the comic convention according to which the awakening of love causes the male to lapse into trances and to lose control of his body, in particular to be everywhere in danger of falling down or of breaking things. *The Lady Eve* contains, as we saw, another virtuoso treatment of this convention. And even Spencer Tracy, whom it is hard to humiliate, is asked by the genre to suffer these indignities. Second, it harps upon repetition. Beyond the texture of verbal repetitions and the beginning and ending *tableaux vivants*, and beyond the two "I'll be with you in a minute, Mr. Peabody" exits, and the two kidnappings in stolen cars, and the two scenes of serenade under the second-story windows of respectable houses, and two golf balls and two convertible coupés and two purses and two bones and two bent hats, there is the capping discovery that there are two leopards in Connecticut. My idea, then, is that this structure is to be understood as an interpretation of the genre of remarriage in the following way: the principals accept the underlying perception that marriage requires its own proof, that nothing can show its validity from outside; and its comedy consists in their attempts to understand, perhaps to subvert, to extricate themselves from, the necessity of the initial leap, to move directly into a state of reaffirmation. It is as though their summer night were spent not in falling in love at first or second sight, but in becoming childhood sweethearts, inventing for themselves a shared, lost past, to which they can wish to remain faithful. (Among the other, nonexclusive, perceptions of their final setting, it can be read as a tree house or a crib.) It is a kind of prehistoric reconstruction. That this must fail is not exactly funny. Grant, in particular, never smiles.

The concluding tableau is a repetition, or interpretation, not alone of the opening shot of Grant, but of the image upon which the final scene (preceding the Epilogue) had closed. There Grant faces the second leopard, the wild one, the killer, using correctly this time an appropriate implement, a tamer's tool before him, and coaxes the beast into a cage, or rather a cell; it is, as it happens, the particular cell in which Hepburn had been locked. In this final game (playing tamer and rescuer), the woman is now standing behind the man, and, after their victory, he turns to face her, tries to say something, and then loses consciousness, collapsing full-length into her arms for their initial embrace. Somewhat to our surprise, she easily bears his whole weight. Nature, as in comedies it must, has taken its course.

This sub-conclusion is built upon a kind of cinematic, or grammatical, joke. The cutting in this passage back and forth between the leopards emphasizes that we are never shown the leopards within the same frame. It thus acknowledges that while in this narrative fiction there are two leopards, in cinematic fact there is only one; one Baby with two natures; call them tame and wild, or call them latent and aroused. It is this knowledge, and acknowledgment, that brings a man in a trance of innocence to show his acquisition of consciousness by summoning the courage to let it collapse.

COMMON TO SOME who like and some who dislike *Bringing Up Baby* is an idea that the film is some kind of farce. (It would be hard to deny that some concept of the farcical will be called upon in dealing with the humor in marriage.) But if the home of this concept of farce lies, say, in certain achievements of nineteenth-century French theater, then, as in other cases, this concept is undefined for film. I do not deny that such achievements are a source of such films, but this merely asks us to think what a source is and why and how and by what it is tapped. Nor would I put it past Howard Hawks, or those whose work he directed, to be alluding in their title to, even providing a Feydeauian translation of, Feydeau's *On purge Bébé*. This would solve nothing, but it might suggest the following line of questioning: Why, and how, and by what, is such a source tapped in this film since neither the treatment of dialogue nor of character nor of space nor of the themes of sexuality and marriage in *Bringing Up Baby* are what they are in Feydeau?

One line of response might undertake to show that this question encodes its own answer, that *Bringing Up Baby* is what it is precisely in negating Feydeauian treatments. This would presumably imply a negation or redemption of (this species of) farce itself, that is, an incorporation, or sublation, of the bondage in marriage into a new romanticizing of marriage.—Would an implied criticism of society be smaller in the latter than in the former case? Not if one lodges one's criticisms of society irreducibly, if not exclusively, from within a demand for open happiness. Feydeauian comedy cedes this demand on behalf of its characters; Hawksian comedy, through its characters' struggles for consciousness, remembers that a society is crazy which cedes it, that the open pursuit of happiness is a standing test, or threat, to every social order. (Feydeau and Hawks are as distant conceptually as the Catholic and the Protestant interpretations of the institution of marriage, hence of the function of adultery.)

What is it about film that could allow the "negation" of theatrical "treatments"? Take the treatment of character, and film's natural tendency to give precedence to the actor over his or her character. This precedence is acknowledged in the capping repetition of the line—the curtain-line for each of the first two scenes—"I'll be with you in a minute, Mr. Peabody." It scans and repeats like the refrain of a risqué London music hall ballad, of course to be sung by a woman. This contributes to an environment for our response to the *expertness* of the pair's walk-off through the revolving door of the restaurant. (That they are as on a stage is confirmed by the inset cut, in mid-walk, to a tracking shot past the astonished Mr. Peabody, who takes the place of an audience.) The authority of this exit, which calls for a bent hat held high in salute in the hand upstage, is manageable only by a human being with Cary Grant's experience and expertise in vaudeville.

As well as in its allusions to, and sources in, farce and vaudeville, this film insists upon the autonomy of its existence as film in its allusions to movies. When I took in Grant's line, in the jail-house scene, "She's making all this up out of old motion pictures," I asked myself, Which ones? (There is a similar jailhouse scene in John Ford's earlier *The Whole Town's Talking*.)* But of course one is invited further to ask oneself why, in so self-conscious a film, Hawks places this allusion as he does. It is a

---

* Andrew Sarris provided this answer at the New York conference at which a version of this reading was presented. I have not seen the Ford film.

line that immediately confesses the nature of movies, or of a certain kind of movie making: the director of the movie is the one who is making all this up out of old motion pictures. (As Hitchcock will incorporate the conclusion of *Bringing Up Baby* into the conclusion of *North by Northwest*, where Grant's powerful hand and wrist save another woman from falling, and we see that the ledge he hauls her onto is his cavebed.) Or: a director makes a certain *kind* of movie; or: a director works within, or works to discover, a maze of kinships. Anyway *this* director does, demanding his inheritance. So Hepburn is characterized by Grant as having or standing for some directorial function. The implication is that the spectator is to work out his or her relation to (the director of) this film in terms of Grant's relation to Hepburn.—So, after all, criticism comes down to a matter of personal attachment! This is why we must adopt some theoretical position toward film!—But I rather imagined that Grant's relation to Hepburn itself might provide a study in personal attachment. At any rate, a theory of criticism will be part of a theory of personal attachment (including a theory of one's attachment to theory, a certain trance in thinking).

I have thus, encouraged by this film, declared my willingness, or commitment, to go back over my reading of it, construed as my expressions of attachment to it. Reconsideration of attachments, and of disaffections, ought to be something meant by education, anyway by adult education, by bringing oneself up. Since for this film I am to proceed in terms proposed by Grant/David's relation to Hepburn/Susan, then before or beyond testing any given form in which I have so far expressed myself about the film, for its accuracy at once to what is there and to what I feel in what is there, I am to ask what I know and do not know about this relation, and what Grant knows and does not know about it. The principal form this question takes for him is, in essence: What am I doing here, that is, how have I got into this relation and why do I stay in it? It is a question all but continually on his mind. So I, as his spectator, am to learn to ask this question about my relation to this film. It will not be enough to say, for example, that I like it, for however necessary this confession may be, that feeling is not apt to sustain the amount of trouble the relation may require, or justify its taking me away from other interests and commitments in order to attend to it. Nor will it be enough to say that I do not like it, should that be required of me, for perhaps I am not very familiar with my likes and dislikes, having over-come them both too often.—If this is a good film, it ought to, if I let it, help teach me how to think about my relation to it.

Earlier, in registering the pace of this narrative as one in which a complete exposition is comically compressed into a stilted prologue, I described the hero as leaving to seek his fortune. His first name for this fortune is, conventionally enough, "a million dollars"; but the first thing he finds on his quest, the first of the nonaccidental accidents which punctuate quests, is a mix-up with an oddly isolated, athletic woman. suddenly appearing from the woods, who looks like a million dollars. (The camera's attraction to Katharine Hepburn's body—its interpretation of her physical sureness as intelligence self-possessed—is satisfied as fully in Cukor's comedies with her as in this of Hawks.) This hero's entanglements with this Artemis from the beginning, and throughout, threaten the award of his imagined fortune, both because she compromises him in the eyes of those who are to award it and because she herself seeks the same million. Yet when at the conclusion she confers it upon him, together with all other treasures, he seems unsatisfied. He gets the money, the lost bone, and the girl, yet he is not happy. What can he, do we think, be thinking of? Why is he still rigid; why is his monstrous erection still false? Do we think: He cannot accept these powers from her, as if these things are her dowry, for in accepting her right to confer them he must accept her authority, her fatherhood of herself? Or do we think: He still cannot think about money any more than he can (or because he cannot) think about sexuality? Or is it: The fate of sexuality and the fate of money are bound together; we will not be free for the one until we are free from the other? Perhaps we shall think, for Luther's reasons, or for Marx's, or Freud's, that money is excrement. I find that I think again, and I claim that such comedies invite us to think again, what it is Nietzsche sees when he speaks of our coming to doubt our right to happiness, to the pursuits of happiness. In the *Genealogy of Morals*, he draws a consequence of this repressed right as the construction of the ascetic ideal, our form of the thinker. He calls for us to have the courage of our sensuality, emblematized for him by Luther's wedding. For this priest to marry, the idea of marriage, as much as that of ordination, is brought into question. I do not say that the genre of remarriage thinks as deeply about the idea of marriage as does, say, the *Pagan Servitude of the Church*. Doubtless our public discourse is not as deep on these matters as it once was. I do say that a structure depicting people looking to remarry inevitably depicts people thinking about the idea of marriage. This is declared by a passage in each of these films in which one or both of the principals try a hand at an abstract theoretical formulation of their predicament. (Among the central members of our genre, *The Awful Truth* contains the most elaborated instance of this, with its concluding philosophical dialogue on sameness and difference, answering to its opening pronouncement about the necessity for faith in marriage.) It is why their conclusions have that special form of inconclusiveness I characterized as aphoristic. Nothing about our lives is more comic than the distance at which we think about them. As to unfinished business, the right to happiness, pictured as the legitimacy of marriage, is a topic that our nation wished to turn to as Hollywood learned to speak—as though our publicly declared right to pursue happiness was not self-evident after all.

About halfway through *Bringing Up Baby*, Grant/David provides himself with an explicit, if provisional, answer to the question how he got and why he stays in his relation with the woman, declaring to her that he will accept no more of her "suggestions" unless she holds a bright object in front of his eyes and twirls it. He is projecting upon her, blaming her for, his sense of entrancement. The conclusion of the film—Howard Hawks's twirling bright object—provides its hero with no better answer, but rather with a position from which to let the question go: in moving toward the closing embrace, he mumbles something like, "Oh my; oh dear; oh well," in other words, I am here, the relation is mine, what I make of it is now part of what I make of my life, I embrace it. But the conclusion of Hawks's object provides me, its spectator and subject, with a little something more, and less: with a declaration that if I am hypnotized by (his) film, rather than awakened, then I am the fool of an unfunny world, which is, and is not, a laughing and fascinating matter; and that the responsibility, either way, is mine.—I embrace it.

4

THE
IMPORTANCE
OF
IMPORTANCE

The Philadelphia Story

And is there not some lingering suspicion that the picture of the trio was already a kind of wedding photo?—that somehow, as Edmund madly says in the final moments of King Lear, "I was contracted to them both. Now all three marry in an instant."

ERHAPS the most obvious difference of George Cukor's *The Phil-adelphia Story* (1940) from its companion members in the genre of remarriage is that it has two heroes, two leading men who are honorable and likable enough for their happiness at the end to make us happy. A good reason for this double presence is to allow us, or to force us, to figure that while each of these men seems a fit candidate for the hand of the heroine, while each loves and appreciates her, and she each of them, one of them is chosen by the genre, as it were, as the more perfectly fit. But on what ground? What has Cary Grant got that James Stewart hasn't got? What is the relevant difference between them? One level of answer would be to say that Stewart is of the wrong social class, and that answer is not so much false as obscure, itself in need of explanation. *It Happened One Night*, as said, is an exception to the apparent rule of the genre that a woman may not marry into a class beneath hers, and it is not clear that Stewart in *The Philadelphia Story* might not achieve exemption on the same ground as Clark Gable had earlier, that of having performed a remarkable and daring feat on the basis of which the heroine is free to give herself to him. In the present case the feat was to have gotten her drunk and then not to have taken advantage of her condition, which proves to establish the context for a particular transformation of her perception. Moreover, Tracy (Katharine Hepburn) has already seen that she and Mike (James Stewart) are like one another, as she finds on reading his book of stories in the town library, saying to him that she knows quite a lot about hiding an inner vulnerability under a tough exterior. But what C. K. Dexter Haven (Cary Grant) has over Mike Macauley Conner is that he and Tracy Lord, as he puts it in a kind of displaced Prologue to the film, in the offices of *Spy* magazine,

135

grew up together. Sidney Kidd (Henry Daniell), the publisher of *Spy*, is introducing Dexter to Mike and to Mike's steady friend Liz Imbrie (Ruth Hussey), who work, respectively, as writer and photographer for Kidd, though each has better things in mind. Dexter will introduce them, Kidd says, without saying why, into the Lord household for the weekend festivities so that they can get their behind-the-scenes coverage of Tracy's marriage to George Kittredge (John Howard)—"that man of the people," "Presidential timber"—focusing on the difficult and private phenomenon called Tracy Lord, a coverage *Spy* magazine will feature as—Kidd searches his imagination for a lead—The Philadelphia Story. In this setting Dexter's words about his and Tracy's having grown up together are meant ironically, to refer to their marriage and divorce; but Tracy's mother uses just these words the next day to state what we do not doubt is the literal truth. Having grown up together, or anyway having in some way created a childhood past together, remains a law for the happiness of the pair in the universe of remarriage comedies. Mike's presence confirms this law while at the same time it establishes that what makes George unfit for Tracy is not the sheer fact of his emergence from a lower class.

Stewart/Mike's role goes beyond these defining functions. He is essential to the way this narrative modifies the structure in which the woman is re-won, won back. This comes out in the interview he precipitates with C. K. Dexter Haven, leaving the party at Uncle Willie's the night before the wedding to seek him out at home. After some desultory talk, he comes out with, "Doggone it C. K. Dexter Haven, either I'm going to sock you or you're going to sock me." This creates an intimacy between them which leads to the plan to counter Sidney Kidd's threat of blackmail. But more significantly Mike has uttered a prophecy which is fulfilled two sequences later when Dexter does indeed sock Mike on the jaw. The point of the prophecy is that for all their identifications with Tracy and for all their shared knowledge that George is not the man for her, and for all their lecturing of the woman, one or the other of them is going to have to *claim* her, to risk declaring himself as her suitor, and specifically to claim her from the other. That Mike must claim her from Dexter seems obvious enough to us, as it does the next day to Tracy when she directs to Dexter her confession of her fears about herself. That Dexter must claim her from Mike is not yet clear at the moment he utters the prophecy (perhaps it is part of the prophecy) but it is manifest as Mike confronts Dexter and George on the terrace carrying the limp Tracy in his arms. He has elicited an old expression of desire from her. Dexter knows, George does not, what that desire is; we witnessed his upholding its importance to her the preceding afternoon when he brought her the model of the *True Love*, the boat Dexter designed and built for their honeymoon. So Dexter has to claim specifically that that desire of hers is what he desires, and even that it is by rights directed to him. We will come back to this encounter.

Who is this man, C. K. Dexter Haven/Cary Grant? When George, furious and confused at Tracy's refusal, or rather acceptance, of his suggestion to let bygones be bygones, turns on Haven with the accusation, "Somehow I feel you had more to do with this than anyone," and Dexter replies, "Maybe, but you were a great help," we laugh both at the victory of light over darkness and also at the truth, hard to locate, of Dexter's power, apparently some mysterious power to control events. The magic invoked by the genre seems localized in this figure. Surely this has to do with his sheer physical attractiveness; he is, after all, or before all, Cary Grant. But our genre leads us to suspect that it also has to do with something of paternal authority in him.

In speaking earlier of the genre's emphasis on the father-daughter relationship, and adding to this the notable absence of the woman's mother as something that compounds that emphasis, I noted that *The Philadelphia Story* is an exception which proves this rule of absence. This father dresses down his daughter in a long aria—in lines like none other given a father in our films—beautifully delivered by an actor of stature, which contains words difficult to tolerate (like "A husband's philandering has nothing to do with his wife") but which ends with a couple of blows so telling as to lend to the whole speech a coherent gravity (". . . you lack the one thing needful, an understanding heart . . . What's more, you've been speaking like a jealous woman"). The father, in returning, would require a showdown with Tracy since she has been taking over the position of head of the household (encouraged her mother to turn away from her father; refused to invite him to her wedding; long ago decided that her sister's name should not be Diana). But I take it as essential to his aria that it occurs *in the presence of the mother*, as a kind of reclaiming of her from Tracy. The mother's acceptance of his words has two effects: it demonstrates to Tracy that her mother cannot or will not offer her protection or comfort or guidance, and there is next to no further exchange between them in the film (Tracy does say later that night at the party, "Oh Mother. I thought you went home hours ago"); and it seems to deprive the mother of her mental competence, so that while she continues to be present, her mind is absent (she is puzzled the next day about who Mike is, and then after apparently recognizing him asks him if he has a violin string). Something dire has happened to the woman who had had a tender, intimate moment with Dexter the morning before and to whom he had said, "That's the old Quaker spirit, Mother Lord!"

This connection between the woman's mother and the woman's first husband, in the light of the mother's eventual alteration (and also thinking of the daughter's consoling her mother's expression of loneliness by saying in one of her early lines, "We just chose the wrong first husbands, that's all") prompts me to cite a passage or two from Freud's 1931 essay entitled "Female Sexuality": "the phase of exclusive attachment to the mother, which may be called the pre-Oedipal phase, possesses a far greater importance in women than it can have in men . . . Long ago . . . we noticed that many women who have chosen their husband on the model of their father, or have put him in their father's place, nevertheless repeat toward him, in their married life, their bad relations with their mother . . . Perhaps the real fact is that attachment to the mother is bound to perish, precisely because it was the first and was so intense; just as one can often see happen in the first marriages of young women which they have entered into when they were most passionately in love. In both situations the attitude of love probably comes to grief from the disappointments that are unavoidable and from the accumulation of occasions for aggression. As a rule, second marriages turn out much better."* The idea is that the bad relation with the mother may be shed along with the bad first marriage. I suppose this is not something Freud is recommending as a general solution to a psychological impasse, but his show of worldly wisdom here is, as is characteristic with him, a response to a problem for which he sees no solution and claims to find no fully satisfactory explanation: "When we survey the whole range of motives for turning away from the mother

---

* Standard ed. 21, pp. 230–231, 234.

which analysis brings to light . . . all these motives seem nevertheless insufficient to justify the girl's final hostility."* (Astoundingly, to me, Freud does not consider that the girl may be responding to a hostility directed to her by her mother. The unmasking friend of the child's sexuality seems at this moment sentimental about mother love.) The romantic insistence on the woman's mutual happiness with her father might accordingly be seen as a wish for refuge from an earlier, apparently insoluble conflict with her mother. And the moral of the genre of remarriage might be formulated so as to include the observation that even good first marriages have to be shed; in happy circumstances they are able to shed themselves, in their own favor.

Dexter's authoritativeness, or charisma, may poorly or prejudicially be interpreted as a power to control events. Maybe it is a power not to interfere in them but rather to let them happen. (The association of an explicitly magical person with a power of letting others find their way, where the others are children and the person in question is a teacher, is given one permanent realization in Jean Vigo's *Zero for Conduct.*) Dexter's refusal to interfere with events, anyway with people's interpretations of events (as if always aware that a liberating interpretation must be arrived at for oneself) is expressed in his typical response to those who offer interpretations of *him,* either to toss their words back to them (George: "I suppose you pretend not to believe it?" Dexter: "Yes, I pretend not to"); or to use his characteristic two- or three-syllable invitation to his accusers to think again, asking "Do I?" (have a lot of cheek); "Wasn't I?" (at the party); or "Am I, Red?" (namely, loving the invasion of her privacy). He is a true therapist of some sort.

This magus can readily also be understood as a figure serving as surrogate for the film's director—a function played in *Bringing Up Baby* and in *The Lady Eve,* as noted, by the woman of the central pair, and in *It Happened One Night* and *His Girl Friday* by the man. Here it is to the point that while Sidney Kidd commissions the story and hires the writer and photographer, serving so to speak as a producer, it is Dexter who puts them into the picture. It is this directorial power that George is vaguely responding to when he accuses him of manipulating the ending they have come to; and Dexter openly directs, or casts and costumes and writes, the ensuing wedding ceremony. The most explicit

* Ibid., p. 234.

statement of this function in the dialogue is the exchange that runs from George's pompous "A man wants his wife to . . ." through Tracy's leaping in as if to cover George's vulgarity ". . . to behave herself. Naturally," to Dexter's comment, "To behave herself naturally." This gratifying re-emphasis is a piece of instruction at once moral and aesthetic—it speaks of a right way to live but at the same time tells how to act in front of a camera, and specifically how to deliver a line.

This climactic, simultaneous advice to character and to actress is something to be expected if I am right that the subject of the genre of remarriage is well described as the creation of the woman, or of the new woman, or the new creation of the human. For this description is meant at once to characterize an emphasis taken by the narrative on the question of the heroine's identity and an emphasis taken by the cinematic medium on physical presence, that is, the photographic presence, of the real actress playing this part, an emphasis that demands, without exception, some occasion for displaying or suggesting the naked body of the woman to the extent the Production Code will allow. Thus does film, in the genre under consideration, declare its participation in the creation of the woman, a declaration that its appetite for presenting a certain kind of woman a certain way on the screen—its power, or its fate, to determine what becomes of these women on film—is what permits the realization of these narrative structures as among the highest achievements in the art of film. This is something I have meant by suggesting that in the genre of remarriage film has found one of its great subjects.

In *The Philadelphia Story* the narrative emphasis on identity takes the form of the question whether the heroine is a goddess made of stone or of bronze, or whether a woman of flesh and blood; and its cinematic occasions for studying Katharine Hepburn's body take their cue from the presence of water, first watching her trained dive into her swimming pool, and second, in the moment mentioned earlier that leads up to the crisis of the sock on the jaw, sensing her weight and her pliancy as James Stewart enters carrying her in a bathrobe falling open at the knees, singing a triumphant "Over the Rainbow," a beautiful song about how dreams come true. Citing the form of Old Comedy as one in which the heroine may undergo something like death and revival, and noting that we can understand this entire narrative as one tracing the death and revival of the woman's capacity to feel, her rebirth as human (or, as D. H. Lawrence more or less put the matter, the dead spirit re-surrected as body—and Lawrence scarcely lectured his heroines more relentlessly on this topic than Tracy is lectured by the four men in her life), we will hardly avoid seeing the carrying posture, if only in retro-spect, as symbolic of her death as goddess and rebirth as human. But just as significantly, the posture is an inherently ambiguous one. Begin-ning with the form of rescue from water, it alludes to the posture of a father carrying a sleeping or a hurt child, or the gesture of a husband carrying a bride across a threshold. Dexter has a moment of concern about it: "Is she all right?", at which point Tracy raises her head and mutters darkly, "Not wounded sire, but dead." Each of the characters present in fact interprets the gesture, puts his or her imagination to work on it. George's interpretation, as he will say the next day, didn't take much imagination, to which Tracy will answer, "No. Just an imagi-nation of a particular kind." Dinah has perhaps a similar interpretation; Mike has another, he speaks of the wine hitting her as she hit the water. It is a question whether Dexter has a competing interpretation exactly; it seems essential, rather, that his guiding interest is in waiting to see what Tracy's interpretation will be, which comes to seeing how and whether she will remember the event.

The moment is in any case a crucial one, shown on its surface by its being the only shot in the film in which all and only these four are framed together, the woman and her three suitors (the right combina-tion for a fairy tale). And in some ways it is the most comic moment in the film, prepared for by Dexter's trying to get George to leave before he sees what's coming; intensified by Mike's stopping singing and then, heroically, starting to sing *again*, in full consciousness of the situation; and capped by Tracy's threefold greeting: "Hello, Dexter; Hullo, George; Hallo, Mikey." But this is also one of the two most anxious moments, posing inescapably the question of tomorrow for the woman, of what she is going to do. It isolates the fact that even where there is a festival ahead, it marks the exercise of choice and of change, and the choice and change may be painful, as painful as becoming created, be-coming the one you are, and as becoming one in marriage.

It is part of our understanding of our world, and of what constitutes an historical event for this world, that Luther redefined the world in getting married, and Henry the Eighth—one of the last figures Shake-speare was moved to write about—in getting divorced. It has since then been a more or less open secret in our world that we do not know what legitimizes either divorce or marriage. Our genre emphasizes the mystery of marriage by finding that neither law nor sexuality (nor, by implication, progeny) is sufficient to ensure true marriage and suggesting that what provides legitimacy is the mutual willingness for remarriage, for a sort of continuous reaffirmation, and one in which the couple's isolation from the rest of society is generally marked; they form as it were a world elsewhere. The spirit of comedy in these films depends on our willingness to entertain the possibility of such a world, one in which good dreams come true.

There are specific precedents in Shakespearean romance for a structure which puts together an inaccessibility to normal society with a peculiar curse, and beauty, of imagination, in which a wife is accused of a particularly vulgar faithlessness (on the basis of the evidence of the senses) and in which she is perceived as made of stone. This is *The Winter's Tale*, and it entails its companion piece *Othello*, which I have interpreted as, so to speak, a failed comedy of remarriage, a narrative in which the reunion is hideously parodied and becomes possible only a moment too late. The three males of *The Philadelphia Story* may be construed as dividing up Othello's qualities—Dexter taking up his capacity of authority, Mike his powers of poetry and passion (Hepburn insists that Mike's stories are poetry and Mike heartily agrees), George his openness to suspicion and jealousy. Such a division simplifies the problem of character and it makes more manageable the obligation of romantic comedy to expel jealousy and envy in preparation for a happy ending. *The Winter's Tale* also harps on the idea of dreaming, but it is *A Midsummer Night's Dream* that more closely anticipates the conjunction of dreaming and waking, and of apparent fickleness, disgust, jealousy, compacted of imagination, with a collision of social classes and the presence of the whole of society at a concluding wedding ceremony, a presence unique among the members of our genre in *The Philadelphia Story*.

And *Midsummer Night's Dream* is built from the idea that the public world of day cannot resolve its conflicts apart from resolutions in the private forces of night. For us mortals, fools of finitude, this therapy must occur by way of remembering something, awakening to something, and by forgetting something, awakening from something. In the

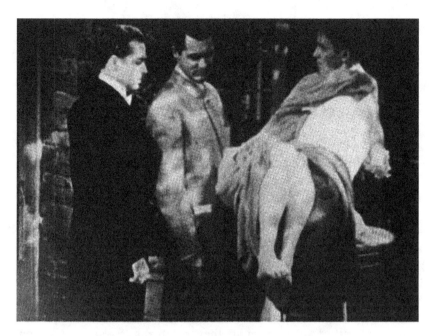

We will hardly avoid seeing the carrying posture as
symbolic of her death as goddess and rebirth as human.
But just as significantly, the posture is an inherently
ambiguous one.

language of *The Philadelphia Story* this is called getting your eyes opened,
and the passage both to the private forces of night and to the public
world of consequences may be accomplished by champagne, or some
other concoction of liquors and juices. Dexter offers Tracy a stinger,
made he says "with the juice of a few flowers." In *Midsummer Night's
Dream* the eyes are analogously closed and opened by what it calls the
liquor or juice of certain flowers or herbs, used externally. It is upon
such application that Titania becomes enamored of an ass. Tracy presumably became enamored of an ass by the more up-to-date agency of
what we call "the rebound," what Dexter calls "a swing"; but she
wishes to do with her creature what Titania wishes to do with hers, to
"purge [his] mortal grossness so, / That [he shall] like an airy spirit
go." Tracy is shown to try purging, or anyway covering, George's

lower-class grossness so that he can go like an airy aristocrat on horse-back (or rather to cover his failed attempt to cover his grossness) by rubbing dirt into his new riding habit.

With *Midsummer Night's Dream* as subtext, other moments in *The Philadelphia Story* find a special comprehensibility. Take, for instance, the exchange between Tracy and Mike as they meet the morning after. Tracy says something about the handsomeness of the day and Mike retorts, "Yeah. What did it set you back?" to which she answers, "Nothing. I got it for being a good girl." (Among Dinah's first words were that it won't rain because Tracy won't stand for it.) That a beautiful midsummer day is something Tracy owns recalls Titania's proof that she is a spirit of no common rank: "The summer still doth tend upon my state." Take again the peculiar character of Uncle Willie. I accept his presence as sufficiently justified by his permitting Tracy's line "What has class to do with it? Mack the night-watchman is a prince among men, Uncle Willie is a . . . pincher." But how does he get his specific budget of characteristics? If you let yourself be puzzled by the image of Dinah and Uncle Willie riding together through a forest in a pony-cart, as if creatures from another realm, and if you speculate on the fairy realm of *Midsummer Night's Dream*, then when Uncle Willie says his head just fell off, you might think of the predicament of Bottom and his temporary head. There is confirmation of this thought in considering that Bottom is the name of what it is of which Uncle Willie is the pincher. (Certainly the pun between Bottom's name and his temporary ass's head is no less blatant in Shakespeare.) I am willing to go further than this and see in Tracy's wafting perfume from her riding kerchief behind the back of her preoccupied uncle, who is spying into *Spy* magazine, a kind of memory of Titania's instructions to her elves on how to treat her gentle mortal: "pluck the wings from painted butterflies, / To fan the moonbeams from his sleeping eyes." Uncle Willie's last words are "Peace, it's wonderful," and the last words of Oberon's speech which predicts the end of the play are: "all things shall be peace."

I am not interested to try to provide solider evidence for the relation of *The Philadelphia Story* and *A Midsummer Night's Dream*. I might rather describe my interest as one of discovering, given the thought of this relation, what the consequences of it might be. This is a matter not so much of assigning significance to certain events of the drama as it is of isolating and relating the events for which significance needs to be assigned. It would not satisfy my curiosity to reduce the problems of *Philadelphia Story* to those of *Midsummer Night's Dream*, because my curiosity is exactly as strong to understand why the concerns of *Midsummer Night's Dream* have worked themselves out in their particular shapes. This will first require learning what these "concerns" are and how to think about those "shapes."

But granted some more or less specific relation to Shakespeare's romantic comedy, does it help to think of C. K. Dexter Haven as Oberon? The bare possibility of the question brings out the fact of Dexter's quality of authority, unmistakable if intangible, as something to which criticism must assign significance. I mention in passing that Oberon is invisible to mortals, as is the figure of the film director for whom, as I have claimed, Dexter, among other things, is a surrogate. I have also acknowledged that Dexter is more literally the magical Cary Grant. But who is Cary Grant? I mean, what becomes of this mortal on film?

It seems to me that George Cukor is calling upon the quality of Grant's photogenesis discovered, as I suggested earlier, in the comedies Grant made with Howard Hawks—I mean the air he can convey of mental preoccupation, of a continuous thoughtfulness that makes him spiritually inaccessible to those around him. This quality of the sage gives to his privacy, his aliveness to himself, a certainty and a depth. We know about Dexter that his wife divorced him because of his drinking, which she claims made him so unattractive (a phrase that serves to focus attention on how attractive this man is). He calls this problem of his "my gorgeous thirst." What is this thirst, which Tracy could not tolerate, a thirst for? And in curing himself of his thirst for alcohol, has he, are we to understand, cured himself of his gorgeous thirst? If it was for the same thing Clark Gable was hungry for in *It Happened One Night*, we might call this thing love, understood as imagining someone hungry for the same things you are yourself hungry for. (It is my claim that hunger in that film is equated with imagination.) Since Dexter's praise of alcohol lies in its capacity to open your eyes to yourself, we might think of his thirst as for truth, or for self-knowledge, as well as for her desire, since his implied rebuke to her (that her eyes are closed to her own desire) is that what she could not bear was his thirst for whatever it is the alcohol represented, call this their marriage. He seems pretty clearly, and unapologetically, to be thinking about it still,

still thirsting. (His curing himself of his substitute addiction, and more-over curing himself by reading, by an absorption in art, is understand-able as the act of self-mastery that has lent him his special powers.)

Then how does he conceive the cause of the end of the marriage whose thread he wishes to pick up again? When Tracy points out to him that drinking was *his* problem he replies, "Granted. But you took on that problem when you married me. You were no helpmeet there, Red. You were a scold." This, however, is once more exactly a brief for his divorce from her, based on Milton's understanding of God's deci-sion to "make him [Adam] an help meet for him" as the perception that "a meet and happy conversation is the chiefest and noblest end of mar-riage." The conjunction of being a helpmeet with being willing to con-verse, a contrary of being a scold, comes out again in a late exchange between Tracy and Dexter as she refuses an offer of a drink from him, warns him never to sell the *True Love*, tells him she'll never forget that he tried to put her back on her feet today, and then collapses on the re-mark, "Oh Dext, I'm such an unholy mess of a girl," to which he re-sponds, "Why that's no good, that's not even conversation."

In adducing Milton's view of the matter of conducting a meet and happy conversation, I have emphasized that while Milton has in view an entire mode of association, a form of life, he does also mean a capac-ity, say a thirst, for talk. And I do not know any words on film that seem to satisfy better the thirst for conversation than those exhibited by these Hollywood talkies of the thirties and forties. Talking together is for us the pair's essential way of being together, a pair for whom, to re-peat, being together is more important than whatever it is they do to-gether.

Because I am working with a notion of a genre that demands that a feature found in one of its members must be found in all, or some equivalent or compensation found in each, I am bound to ask what happens to the fact that *The Philadelphia Story* is the only film among the members of our genre in which the pair's happiness is refound, ap-parently, in the larger world in which they divorced, literally in the place they grew up together, not in removing themselves to a world apart from the public world, a world of their own making, of adventure. This odd feature would reach a satisfactory equivalence in *The Philadel-phia Story* if its pair could be understood to regard their own larger society as itself world enough elsewhere, itself the scene of adventure. What could that mean?

It could mean, for example, that they understand their marriage as exemplifying or symbolizing their society at large, quite as if they are its royalty; and their society as itself embarked on some adventure. George is confusedly thinking something more or less like this when he declares toward the end that his and Tracy's marriage will be of "national importance." And Tracy had toward the beginning defended George to Dexter by claiming that he is already of national importance, in response to which Dexter winces and says she sounds like *Spy* magazine. Yet George and Tracy may be wrong not in the concept of importance but in their application of the concept. What George had said was that Sidney Kidd's presence gives their wedding national importance, and this leads George to put aside his doubts about the woman he is involved with and go ahead with the ceremony. It is to this that Tracy finally says, "And goodbye, George."

George lives his life outside in, so he is never free from the idea that something external to his life can give it importance. His twin assumption is that Sidney Kidd makes the news rather than has a nose for it, as if the public's news and publicity were one and the same thing, which really amounts to saying that nothing is news any longer, an idea that we have in recent decades become increasingly tempted by. Mike's and Dexter's happier thought is that when Sidney Kidd makes news this is a scandal. George's view, from outside in, is not exactly what Tracy most despises; it is doubtful whether she can so much as conceive of it. What primarily motivates her is rather the *fear* of living inside out, of being exposed. This is why her despising of publicity is rather too strong for one of normal democratic tolerances.

But then why *is* Sidney Kidd present?—present, I mean, at the wedding. He had come to the Lord house because Dexter had sent for him to read the counter-blackmail story he composed from Mike's facts. Reading the threatened revelations about himself, Kidd gives up the idea of using the threatened revelations about Tracy's father to gain coverage of Tracy's wedding. This much we know from a brief message Tracy's mother delivers, distractedly, to Dexter: "A Mr. Kidd says to tell you that he's licked, whatever that means." My question is why Kidd then hangs around for the wedding. Of course we are invited to think, seeing him suddenly snapping pictures at the end, that while his plan to insert his lackeys failed, *he*, never missing an opportunity, gets the story himself. But why should we care about that? This way of accounting for Kidd's presence at the wedding leaves out the two interesting facts about it: that it is not the same wedding as the one Kidd had elaborately arranged blackmail to get at; and that he assumes, correctly it turns out, that he is going to be allowed to walk away with his pictures of this wedding. These are facts that show Kidd to be who he is not because he has merely the power to get the news but because he has a nose for it. I think we must understand Kidd's presence, accordingly, to be a signal that it is after all *this* wedding, this remarriage, that is of national importance. (No doubt the bridegroom seems out of costume for it. But then, as Thoreau put it, "Beware of enterprises that require a change of clothes.")

"Importance" is an important word for Dexter, and throughout the film. In his main lecture to Tracy, the one in which he accuses her of having been a scold, he recurs to a sore point between them, her failure to remember the night she got drunk and stood naked on the roof wailing like a banshee, a failure he links to her inability to tolerate human weakness, imperfection. And when she counters with, "You made such a fuss about that silly incident," he takes the point as far home as it is going to get in the words of this film: "It's enormously important . . . You'll never be a first-class person or a first-class woman until you . . ." do something like accept human imperfection, frailty, in others and hence in yourself. For us, bearing in mind the images of the woman at night that we are given to glimpse, this imperfection, this lack of something, this want of something, is desire. Dexter is saying that her condemnation to being divine, worshiped instead of loved, is her ignorance of her sexuality, her demand to remain a goddess "intact." He calls her chaste, virginal, upon which she furiously returns, "Stop using those foul words."

I have said before that the idea of innocence, indispensable to classical romance as a preoccupation with virginity, remains at issue in the genre of remarriage, where the status of literal or physical virginity is presumably no longer a question. The blatant preoccupation in *The Philadelphia Story* with literal virginity, anyway with purity as chastity, is unique among the comedies of remarriage. It extends from Dexter's painful accusations of Tracy to the effect that she is hanging on to her virginity, through the associated imagery of her as a goddess, and concludes with one of the concluding lines of the film, as she invites Liz to be her maid of honor and Liz replies, "Matron. Remember Joe Smith." Liz's easy clarity about her condition underscores Tracy's perplexity in discovering how to shed her virginity.

Freud had also been moved in "The Taboo of Virginity," a dozen years earlier than "Female Sexuality," to voice his impression that "second marriages so often turn out better than first." Earlier in the essay on virginity he had said, "The husband is almost always so to speak only a substitute, never the right man," thus invoking the principle that the finding of an object is in fact the refinding of an object. But is the second marriage better because the second husband is mysteriously spared the status of substitute, of being the wrong one? In this earlier essay, Freud relates the superiority of the second marriage to "the paradoxical reaction of women to defloration," namely that it both binds them lastingly to the man who first acquaints them with the sexual act, but also "unleashes an archaic reaction of hostility toward him." Of all the strategies Freud cites for avoiding the consequences of what he calls this paradoxical reaction, none seems to me as neat or as satisfactory as the idea of remarriage, according to which you are enabled to remain with the one to whom you have been bound, by discharging your hostility on a past life with that one, or with a past version of that one. Two cautionary remarks about this idea. First, we are by now clearly speaking in a psychological mode, so that I am not talking about physical intactness but rather I am supposing that there is such a thing as psychological or spiritual virginity, something for which physical virginity is a trope; and that there is such a thing as psychological or spiritual defloration which may be imagined to have the paradoxical consequences of binding and hostility that Freud perceived. Second, I mean to commit myself to the attempt to think through the consequences of the Blakean concept of spiritual virginity in place of, or as an interpretation of, the notion Freud uses to describe the woman's unhappiness in her first marriage, namely frigidity, in any case a notion to be suspicious of, and not, it seems to me, exactly what Dexter is accusing Tracy of. (While I have no intention of attempting to describe a sexual life for Tracy that we are given insufficient evidence to get very far with, I would not have us overlook certain facts we do have and which must go into what we think this woman is. What do we make of her wafting the perfume at her problematic uncle, "playing with fire" as he calls it? It is a childish trick, entered into, it appears, as a sort of bond with Dinah, as it were on a dare. It strikes me as a sexual prank performed as if within safely narcissistic or still innocently incestuous precincts. And what do we think of her having changed her sister's name from Diana to Dinah? Was this because she felt the name of the goddess of chastity belonged to *her*? Or that her sister was hers to name? "No doubt this sounds quite absurd, but perhaps that is only because it sounds so unfamiliar."* And what do we imagine she makes of the fact that she does not remember whether she did or did not have a sexual encounter the night before? Since she has the revelation that men are wonderful on learning that a man did not take advantage of her, she has apparently harbored the idea that (male) sexuality is inherently a matter of taking advantage, an idea Dexter's gorgeous thirst did not succeed in refuting for her.)

Dexter's demand to determine for himself what is truly important and what is not is a claim to the status of a philosopher. George's acceptance of the status of importance is the mark of one antithetical to the work of philosophy, the mark of the unexamined life. But is what Dexter claims to be enormously important, a matter of one's most personal existence, to be understood as of national importance? How is the acceptance of individual desire, of this form of self-knowledge, of importance to the nation?

These questions take me back to Milton's tract on divorce. Even before knowing what specific ground there may be for divorce we know that there must be such a ground, because

> He who marries, intends as little to conspire his own ruin as he that swears allegiance; and as a whole people is in proportion to an ill government, so is one man to an ill marriage. If they, against any authority, covenant, or statute, may by the sovereign edict of charity save not only their lives but honest liberties from unworthy bondage, as well may he against any private covenant, which he never entered to his mischief, redeem himself from unsupportable disturbances to honest peace and just contentment. And much the rather . . . For no effect of tyranny can sit more heavy on the commonwealth than this household unhappiness on the family. And farewell all hope of true reformation in the state, while such

* "Female Sexuality," p. 239.

an evil as this lies undiscerned or unregarded in the house: on the redress whereof depends not only the spiritful and orderly life of our grown men, but the willing and careful education of our children.

Since nothing can be of greater moment to the state than combating the effect of tyranny, nor in general than reformation, nothing can be of greater moment to it than freedom from an unhappy marriage. It follows that the state at large has no solace or interest sufficient to revive one in bondage to that unhappiness. It seems to follow more specifically from Milton's descriptions that one who suffers its effects makes the commonwealth suffer in terms very like those in which he himself suffers. The unhappiness in marriage, remember, is bondage to "a mute and spiritless mate"; now we are told that its effect on the commonwealth is a "heaviness," and that without redress from it the life of its members cannot be "spiritful and orderly," that is, life in the commonwealth is dispirited and disorderly, or anarchic. It is as if the commonwealth were entitled to a divorce from such a member, but since from a commonwealth divorce would mean exile, and since mere unhappiness is hardly grounds for exiling someone, the commonwealth is entitled to grant the individual divorce, hoping thereby at any rate to divorce itself from this individual's unhappiness. It seems to me accordingly to be implied that a certain happiness, anyway a certain spirited and orderly participation, is owed to the commonwealth by those who have sworn allegiance to it—that if the covenant of marriage is a miniature of the covenant of the commonwealth, then one may be said to owe the commonwealth participation that takes the form of a meet and cheerful conversation.

I understand these film comedies to be participating in such a conversation with their culture. The general issue of the conversation might be formulated this way: granted that we accept the legitimacy of divorce, what is it that constitutes the legitimacy of marriage? If we do not know that a marriage has been effected, how can we know whether there has been a successful divorce, especially when the couple are evidently unable to *feel* themselves divorced, when the conversation between the divorced pair is continuous with the conversation that constituted their marriage? Now the bondage in question is not in these cases to an isolating unhappiness but to an isolating happiness, or to a

In adducing Milton's view of the matter of conducting a meet and happy conversation, I have emphasized that while Milton has in view an entire mode of association, a form of life, he does also mean a capacity, say a thirst, for talk.

shared imagination of happiness which nevertheless produces insufficient actual satisfaction. More specifically, I claim for *It Happened One Night* that the conversation invokes the fantasy of the perfected human community, proposes marriage as our best emblem of this eventual community—not marriage as it is but as it may be—while at the same time it grants, on what may be seen as Kantian and Freudian and Lévi-Straussian grounds, that we cannot know that we are humanly capable of achieving that eventuality, or of so much as achieving a marriage that emblematizes it, since that may itself be achievable only as part of the eventual community. For *The Philadelphia Story*, I am about to claim that its conversation more narrowly focuses such questions on the question of America, on whether America has achieved its new human being, its more perfect union and its domestic tranquility, its new birth of freedom, whether it has been successful in securing the pursuit of happiness, whether it is earning the conversation it demands.

To ENTER MY CLAIM about the conversation in which *The Philadelphia Story* participates, I propose to follow Dexter's advice to Dinah the final morning and to understand what I have seen as a dream—not, as I take his advice, in order to doubt the reality of what I have seen (*that* aspect of its reality, or unreality, is clear enough to me) but in order to look in a certain way for the meaning of those events; and not, as I take him to be advising her, just those of the night before, but of the whole story. This might take me beyond the threat to perceive as *Spy* magazine would have me perceive, as a spectator of obvious scandal, and permit me to acknowledge my participation in these events, or say my implication in them.

The dream I weave—it is more like a daydream—works on a small residue of events and phrases from the film, most of which I have already cited: on the recurrent idea of people coming from different classes, and the repeated notion of being a first-class human being; on the setting of leisure, of luxury, of what Mike calls the privileged class enjoying their privileges; on Tracy's fear of exposure and her responsiveness to Mike; on Dexter's gorgeous thirst and George's expulsion; on a man's wanting his wife to behave herself naturally; on the demand of the genre that the pair are recommitting themselves to an adventure; on Sidney Kidd's being drawn to them as to news of national importance; and on the name Philadelphia.

My dream of the story about Philadelphia is a story about people convening for a covenant in or near Philadelphia and debating the nature and the relation of the classes from which they come. It is not certain who will end up as signatories of the covenant, a principal issue being whether the upper class, call it the aristocracy, is to survive and if so what role it may play in a constitution committed to liberty. The significance of the relation of *The Philadelphia Story* to *A Midsummer Night's Dream* would on this point be the interpretation of aristocrats living in woods on the outskirts of a capital city, as beings inhabiting another realm, a medium of magic, or call it money, which has some mysterious connection with our ordinary lives: we cannot be at peace and clear if they are in conflict and confusion, but it is hard to say whether their turmoil causes ours or ours theirs. And it is very much to the point that Shakespeare's faery realm is the realm of the erotic. (In an essay entitled "The Fate of Pleasure"* that bears variously on our subject, Lionel Trilling remarks that Werner Sombart, in his *Luxury and Capitalism,* "represents luxury as being essentially an expression of eroticism," a representation surely confirmed in the appetite of film, though we might accordingly have to consider further what constitutes luxury on film.)

But the idea of what happened in Philadelphia during the making of our Declaration of Independence and our Constitution is not the whole daydream. It projects further a conversation between the film's preoccupations and some three or four texts or moments in the working out of those covenants in their subsequent two centuries.

No work of serious social criticism was more on the minds of thinking Americans in the period preceding and during the thirties than Veblen's *Theory of the Leisure Class.* One can take *The Philadelphia Story* as a head-on attempt to discredit Veblen's assessment of aristocratic culture as characterized by its conspicuousness and emulation of leisure, of consumption, of waste, which is to say, by an avoidance of productive labor and hence an ignorance of or indifference to the genuine quality of the things in our lives. To the demonstration that Tracy Lord has a horror of the conspicuous, and that one cannot imagine her wishing to emulate anyone, certainly not in what they consume, one might reply on Veblen's behalf in either of two ways: that Tracy's horror of the conspicuous is really a further proof of his point, or that if you are one of the vanishing few who have as much money as the Lords of Philadelphia, and your money is as old, then this theory of leisure may not perfectly apply to you. When Tracy, alone with George for the single time in the film, at poolside after Dexter has departed, having explained to George what "yare" means, cries out, "Oh, to be useful in the world," George responds by saying that he's going to build her an ivory tower and worship her from a distance. This is an especially chilling response exactly because it understands, by denying, both sources of her heartfelt cry, that it is a wish for freedom from her condition of, let me say, enforced or virginal leisure, which she understands as a condition simultaneously of inconsequence or immaturity both in sex and in

* In *Beyond Culture* (New York and London: Harcourt Brace Jovanovich, n.d.).

work. She finds a virginal or narcissistic leisure no longer supportable; but empirical leisure, the kind that has to be chosen for oneself, and that alternates with real work, still maddeningly beyond reach. But however we come to understand how best to join work and love in a satisfying human existence, it seems clear enough to me that Veblen's book is quite deaf to the rights of the sensuous or erotic side of human nature and that it draws too simple, or angry, a picture of the line between the necessary and the luxurious, or the join between the utilitarian and the beautiful.

Tracy's temperament seems better appreciated in the opposite sense of aristocracy appealed to in Tocqueville and in John Stuart Mill. Dreading the tendency in democracy to a despotism of the majority, a tyranny over the mind—another emulation, now of one's neighbor—they looked upon the aristocrat's capacity for independence in thought and conduct, a capacity if need be for eccentricity, as a precious virtue, an aristocratic virtue by which the success of the democratic virtues is to be assessed, to determine whether in its search for individual equality democracy will abandon the task of creating the genuine individual.

This is the sort of thought that enters the third of the texts or moments of conversation concerning our constitution that I find *The Philadelphia Story* to enter, I mean the conversation or fantasy about whether America will produce and recognize in human beings something to call natural aristocracy. Such an idea was classically debated in the correspondence between John Adams and Thomas Jefferson as, late in their illustrious careers, in the second decade of the nineteenth century, they had the leisure to look at their creation and speculate on the conditions necessary for the Union to survive and flourish.* Agreeing on the importance of producing an unprecedented aristocracy, which is just to say, a "rule by the best," they were understandably unable to characterize very satisfactorily what this best is to consist in. The idea seems to affirm that one human being may be better than another and yet to deny (on pain of espousing some repudiated mode of aristocracy) that there is any particular *way* in which one is better, anything one is better *at*. Whatever its unclarity, the idea of natural aristocracy is hard to forget once you have found yourself using it in thinking about America's

---

* A brief, attractive account of the correspondence on this issue is given in William R. Taylor, *Cavalier and Yankee* (New York: Anchor Books, 1963).

aspirations and progress (like the thought of there being a time and place at which the frontier came to an end). I think in fact something like that idea, however dated, even dangerous, it may sound to us, is bound to haunt a society whose idea of itself requires that it repudiate the hierarchies and enforcements of the European past and make a new beginning, to make in effect a reformation of the human condition.

I suggest that this is the idea expressed in Dexter's insistence on what he calls a "first-class human being," an otherwise dark notion. I am prepared to credit his denial that this has to do with a hierarchy of social classes, or with some idea that there are different kinds of human beings, the sort of idea that takes certain others to be primitives or natural inferiors, made for subservience. In the concluding part, Part Four, of *The Claim of Reason*, certain occasions present themselves to me for denying that there are kinds of human beings, and others for allowing that there might be degrees of humanity. Is this evasive? The difference I see in these intuitions may be expressed this way: the natural aristocrat, better in degree but not in kind than his fellows, is not inherently superior to others, possessing qualities inaccessible to others, but, one might say, is more advanced than others, further along a spiritual path anyone might take and everyone can appreciate. (If a talent is something inaccessible to those who do not have it, who are not "gifted" with it, then it would follow that the possession of some talent is not part of the concept of the natural aristocrat.) This is dangerous moral territory. In *The Philadelphia Story* it is under surveillance most explicitly in George's defeat and departure, which we know Dexter has had something to do with. This danger must be run in romance, which wishes the promise of union and renewal, not of expulsion. The drama is lost if this feels merely like a group of snobs ridding themselves of an upstart from a lower class, an inferior. The expulsion is meant, I take it, as a gesture of a promise to be rid of classes as such, and so to be rid of George as one wedded to the thoughts of class division, to the crossing rather than the overcoming of class. A matter of delicate judgment. George's mood fits the thing that has been named *ressentiment*. This is an interpretation of the mood romance names "melancholy" and adopts as its natural foe. At the opening of *A Midsummer Night's Dream*, Theseus' order to his Master of the Revels is: "Awake the pert and nimble spirit of mirth, / Turn melancholy forth to funerals; / The pale companion is not for our pomp."

Given what I just referred to as the moral danger of confusion in this territory, say confusing the cry for justice with a complaint from envy, I take Dexter at the conclusion of *The Philadelphia Story*, when he says to Tracy "I'll risk it. Will you?" to be saying that he'll both risk their failing again to find their happiness together, and also finally risk his concept of that happiness, to find out whether he actually has anything in mind.

The fullest description of what I take him to have in mind is given in, of all things, Matthew Arnold's *Culture and Anarchy*, in which Mill's praise of liberty is contested and from which, I suppose, Veblen got his characterization of the leisure class as the Barbarians (distinguishing these from the middle and lower classes, which Arnold calls the Philistines and the Populace). Arnold also wishes to reconceive the idea of the aristocracy. He wishes to work out the rule of the best to mean the rule of *the best self*, something he understands as existing in each of us. It is of course common not to know of this possibility, but more natures are curious about their best self than one might imagine, and this curiosity Arnold calls the pursuit of perfection. "Natures with this bent," Arnold says, "emerge in all classes . . . and this bent tends to take them out of their class, and to make their distinguishing characteristic not their Barbarianism or their Philistinism, but their *humanity*."

I have elsewhere more than once insisted on the photogenetic power of the camera as giving a natural ascendency to the flesh and blood actor over the character he or she plays in a film, something I take to reverse the relation between actor and character in the theater. I have also spoken of the camera's tendency to create types from individuals, which I go on to characterize as individualities.* Here I recall that long list of actors whose mannerisms or eccentricities so satisfied the appetite of the movie camera during the classical period of film, figures whose distinctness was the staple of impersonators; no self-respecting impersonator could fail to have a Gable and a Grant and a Stewart and a Hepburn routine. This distinctness seems to me a visual equivalent of what Tocqueville and Mill mean by the distinction in aristocracy, by the freedom it projects for itself. It seems to me, further, that there is a visual equivalent or analogue of what Arnold means by distinguishing the best self from the ordinary self and by saying that in the best self

* *The World Viewed*, enlarged ed., pp. 33, 175.

class yields to humanity. He is witnessing a possibility or potential in the human self not normally open to view, or not open to the normal view. Call this one's invisible self; it is what the movie camera would make visible. Perhaps it may discover more than one such self, and not all of them good ones. (It may be making visible what Blake calls our Emanations and our Specters.) I am trying to sketch out a stratum of explanation for the fact, which I cannot doubt, that in these comedies film has found one of its great subjects. I do not say that film is inherently democratic, only that the distinctions enforced by clothes, airs, and reputations in ordinary contexts are quite irrelevant to the distinctions it draws for itself. It is this property of film that allows, say, Fellini to discover in the face of a contemporary Roman butcher the visage of an ancient Emperor.

The originality inspired by the love of the best self Arnold calls genius. So much he might have been confirmed in by Emerson, whom he admired, and by Thoreau, if he read him. But when he goes on to call the best self "right reason" he parts company with American transcendentalism. The rule of the best self is the source of the new authority for which Arnold is seeking, the authority of what he calls culture, of what another might call religion, the answer to our narcissism and anarchy. It was his perception of society's loss of authority over itself, its impotence to authorize the use of force to protect itself from disorder, perhaps from dissolution, that prompted Arnold to write *Culture and Anarchy*. In it he distinguishes two forms of culture or authority, the two historical forces still impelling us on the quest for perfection or salvation; he names them Hebraism and Hellenism. "The governing idea of Hellenism is *spontaneity of consciousness*; that of Hebraism, *strictness of conscience*." The world "ought to be, though it never is, evenly and happily balanced between them." Arnold finds that his moment of history requires a righting of the balance in the direction of spontaneity of consciousness more than it needs further strictness of conscience. The more one ponders what Arnold it driving at, the more one will be willing to say, I claim, that Dexter Hellenizes (as, in their various ways, do Shakespeare and Tocqueville and Mill) while Tracy Hebraizes (as Arnold says all America does, and certainly as Veblen does). Now here is what the marriage in *The Philadelphia Story* comes to, I mean what it fantasizes. It is a proposed marriage or balance between Western culture's two forces of authority, so that American mankind can refind its object, its dedication to a more perfect union, toward the perfected human community, its right to the pursuit of happiness.

It would not surprise me if someone found me, or rather found my daydream, Utopian. But I have not yet said what my waking relation to this daydream is, nor what my implication is in the events of the film.

OUR RELATION to the events of film can only be determined in working through the details of the events of significant films themselves. And specifically, as I never tire of saying, each of the films in the genre of remarriage essentially contains considerations of what it is to view them, to know them. Let me now conclude this reading of *The Philadelphia Story* by calling attention to the events of the ending of the film, which have a peculiar bearing on the issue of viewing.

What you may call its narrative concludes, of course, with the wedding. Or perhaps it really concludes with Sidney Kidd's sudden appearance to photograph this conclusion. But the film goes a step further, ending by showing us the photographs Kidd has taken. His pair of photographs throw into question the status, of form and substance, of everything we have seen. The first photograph is of the trio, Dexter, Tracy, and Mike, startled by, and looking in the direction of, the camera (which camera? Kidd's or Cukor's?). This proves to be a page which turns and gives place to a second photograph, of Dexter and Tracy alone, embracing. We have to understand in this succession of photographs Dexter's at the last moment claiming Tracy from Mike. And is there not some lingering suspicion that the picture of the trio was already a kind of wedding photo?—that somehow, as Edmund madly says in the final moments of *King Lear*, "I was contracted to them both. / Now all three marry in an instant." But further, if they were so determined not to let Sidney Kidd cover the wedding, why, having been startled by the first of the pictures he takes, do the pair allow him to continue, deliberately turn their attention away from the camera and go about their business, quite as if they *wanted* the record made, and quite as if having their picture taken *is* their business?—as if the photograph were the document, or official testimony, that a certain public event has taken place, and that the event is essentially bound up with the achievement of a certain form of public comprehension, of the culture's comprehension of itself, of meet conversation with itself, the achievement, in short, of a form of film comedy.

And how do we understand the provenance of this record, that is, how does it get into our hands? Are these pictures part of the coverage as it appears in *Spy* magazine? Conceivably they are from a wedding photo album. We might take them as production stills. But in any of these cases we are seeing something after the fact, whereas didn't we just now take ourselves to be, as it were, present at the wedding? And what does it mean to say that these final two shots are pictures or photographs? How is the rest of what we have seen different? The rest was every bit as much a function of the photographic. Of course the rest was in motion whereas these are still. But that is the question. What is the difference? This question directs us to think about the ontological status of what we have seen and hence about the mode of our perception. Frye observes that at the conclusion of Shakespearean romance in the theater, we find ourselves subjected to a process in which we somehow move from the position of observer to the position of participant. At the end of *The Philadelphia Story* this process appears to be reversed and we find ourselves awakened from the position of illusory participant to that of observer. But this may itself be illusory. For suppose we find that what has happened to us is that we have substituted for the idea of Tracy as a statue the idea of her and her suitors as photographs, or say traded the goddess for a movie star. Then we are threatened with the very position toward her that George found himself in. Is there a way for us as viewers to escape this position?

In any case, the ambiguous status of these figures and hence of our perceptual state will have the effect of compromising or undermining our efforts to arrive at a conclusion about the narrative. For example, shall we say that the film ends with an embrace, betokening happiness? I would rather say that it ends with a picture of an embrace, something at a remove from what has gone before, hence betokening uncertainty.

Will someone still find that my daydream is not sufficiently undermined by this uncertainty, and still accuse me of Utopianism? Then I might invoke Dexter's reply to George's objection to his, and all of his kind's, sophisticated ideas: "Ain't it awful!"

5

COUNTERFEITING
HAPPINESS

His Girl Friday

A key to this film's placement of its images, or displacement of us before them, is to understand the light it gives us to see by.

IN *The Philadelphia Story* Cary Grant returns; in Howard Hawks's *His Girl Friday* (1940) he is returned to. In both a mystery is explicitly raised about why the return has been made: In the displaced Prologue of *The Philadelphia Story* James Stewart attributes motives of revenge to Grant, who disdains the attribution; in the first full sequence of *His Girl Friday* Hildy Johnson (Rosalind Russell) attributes reasonable motives to herself, but none of them quite sticks. At first she says to Walter Burns (Cary Grant), "I came to tell you not to send me any more telegrams," then toward the end of this long interview she says, "I'm getting married tomorrow. That's what I came here to tell you, but you would start reminiscing." (In *Bringing Up Baby*, Hawks's comedy of two years earlier, Cary Grant says to Katharine Hepburn that he is getting married tomorrow. The information has the same effect on her there, of close-up concentration, that it has on him here.) Evidently Hildy, in *His Girl Friday*, does not know why she has come back to see Walter. I do not say that it is obvious why. If it was merely to tell him something, give him a piece of information, she could have telephoned him or sent him a telegram. And why did she bring Bruce (Ralph Bellamy) along? When Frye remarks that in Old Comedy the woman may undergo something like death and revival he also says that she may bring about the comic resolution. In our films the death and revival, if this is present, is of feeling, it has to happen within the woman, and she cannot, nobody can exactly, *bring* that resolution about. But the woman of *His Girl Friday* can be said to bring about the comic initiation.

Walter seems to know why she has come back. What does he know? He knows that she is not being straightforward with her explanations to him and he knows that she knows from his unending messages to

163

her—by telephone, by telegram, by skywriting—that he wants her back and hence that he will use his endless resources of manipulation to get her back. It must follow for him that she has come back to see him because she wants this of him. But why? If she wants to get back together, why does she not, in return, just say so? Evidently his wanting her back is not enough for her; it does not, in itself, provide a *way* back. There is beyond it something he has to do. Would this be for him to claim her? Not exactly, since he has been asserting whatever claim he is in a position to assert all the time she was getting a divorce. He has to do something like demonstrate that his claim is still in effect, that it is justified—to demonstrate that divorce is not forever, not, so to speak, a sacrament, but only, as he says to her, some words mumbled over you by a judge. "There's something between us that no divorce can come between." Evidently this is their marriage, so evidently it is after all some sort of sacrament. But in order to prove that nothing has come between them he has, so to speak, to arrange for her to free herself from her divorce, to prompt her to divorce herself from it. This seems to be what freedom in marriage requires. It calls for some thought.

In reading *The Philadelphia Story* I called attention to Howard Hawks's discovery of Cary Grant's photogenetic tendency to thoughtfulness, some inner concentration of intellectual energy. In *The Philadelphia Story* this photogenetic possibility is modified into a magus, in *Bringing Up Baby* (and *Monkey Business*) into an absent-minded professor. In *His Girl Friday* it serves the character of a trickster, a familiar figure in the classical history of comedy. The characteristic criteria of Walter's thinking are his drumming or fidgeting fingers and his shifting eyes. I say these characteristics are criteria of his thinking in order to register that they tell us what kind of thinking goes on in him, what it consists of, what modification this character subjects it to. The criteria tell us straight off that his thinking is incessant, compulsive, but let us not be overly confident that we know what he is incessantly thinking about. He makes plans and sets traps often enough, but they are the plans and traps of what is called a newspaper man—they are the expression of his nose for news, which is to say, for a pair of convictions: first that the world at all times presents a false face to its inhabitants, second that under the opportune eruptions of a big story there is a truth behind that face that the right nose can track down.

From the moment Walter sees who is waiting for Hildy outside, really from just before that moment, he knows that what he is to do is to rescue her, or rather arrange for her escape. From what? Not from what Bruce will call "Your chance for happiness, to have the things you've always wanted," which is what Walter roughly, or rather with heavy irony, will describe as a life "full of adventure," namely the life of an insurance salesman's wife, "and in Albany, too." But Walter knows no more about the worth of adventure than Hildy does; he is in no position to weigh the comparative values of adventurousness and, say, insurance. I take it that he is being asked to help her escape not from unhappiness—what Bruce offers her is something she genuinely wants—but from a counterfeit happiness, anyway from something decisively less for her than something else. There is pain in this decision, whichever way she turns; it is no wonder she has become confused.

What will constitute her escape? And let us not, in considering this, be overly confident that Walter's powers of manipulation, and the uses to which he will put them, have no limits. His conduct toward Hildy is guided and limited by two things that he wants back from her in a genuine, unmanipulated state, namely the service of her talent as a writer, and her acknowledgment of her need for him.

The sublime business in Walter's initial encounter with Bruce (first pretending to take an old man to be Bruce, then, apprised of his error, shaking Bruce's umbrella), should again not permanently distract us from the possibility that Walter is putting on a performance not for the sake of its deviousness but for its accuracy; it is directed solely to Hildy and is something only she is in a position to appreciate. When Walter explains his mistake by turning to Hildy to say, "You gave me the impression of a much older man," it is not impossible that she had indeed done so, to someone who knows her as well as Walter does, and who is as good a therapist. By the time, a few moments later, Walter invites them to lunch and the three of them head for the elevator, each of the two of them knows the other knows what each of them wants; and each *wants* the other to know. By the end of the ensuing restaurant sequence they both know what the outcome must be. What neither of them knows is how to arrive at it. As Bruce disappeared into the elevator, Hildy had held back to say privately, under her breath, "You're wasting your time, Walter; won't do you a bit of good"; this already feels as if she is egging him on. And Walter's apparently public and conventional response to her, "No, no, glad to do it," as if she had protested his generosity, privately acknowledges her private, or implicit, appeal to him for help.

An old partner in love showing up apparently unexpectedly with what is interpreted as a request for advice and help with a new love, and one not so much unhappy as falsely happy, is also the opening of *Smiles of a Summer Night*. Like Walter, Desirée may be accused of manipulation, but her plan is to produce a context of illumination in which the one seeking advice comes to recognize his true feelings, especially toward *her*. And like Walter, she begins, after some preliminary remembering, by providing a meal for her tutee, the consequences of which are the rest of the plot. (I mention the Bergman film as the merest glance at one of the paths this book cannot itself follow out, into European film. Another such film obviously invoking the project of remarriage is Renoir's *Rules of the Game* (1939); a less obvious instance is Bresson's *Les Dames du Bois de Boulogne* (1945). It is also pertinent that *Smiles of a Summer Night* alludes to *The Marriage of Figaro* and, by negation, to *Der Rosenkavalier*.)

The restaurant sequence with Walter, Hildy, and Bruce is brilliant and satisfying beyond praise. It takes place in the only comforting environment in a film of claustrophobic sets. But the beauty of the sequence lies in the way it plays for these characters, especially for the central pair. This is essential to the working of the sequence. The setting is made to work as their home, both in its palpable atmosphere of conviviality and in its familiarity to them. It is something rather more than Walter and Hildy's home away from home, since pretty clearly they had had no home at home. They were always other places; that was in brief her grounds for divorce. Accordingly, they are as if entertaining a guest—it is they, not Walter alone, who give this party; just as it is they who are manipulating Bruce, not Walter alone, from the time Hildy makes up a story, naturally for the best reason in the world, to get Bruce to put his money in his hat. ("In your *hat*," she repeats emphatically, joining a line of allusions in the film to the behind, that favorite location of Howard Hawks's, that contains "under her Piazza," "right in the Classified Ads," and a woman with a wart on her named Fanny. These allusions invariably invoke an exclusive fellowship.) Their guest is one whose value they disagree about, but they dispute it within a family agreement—within, I wish to teach us to say, a conversation—of a profundity and complexity the guest cannot begin to fathom. The kicks on the shin Hildy gives Walter under the table are familiar gestures of propriety and intimacy; and the pair communicate not only by way of feet and hand signals but in a lingo and tempo, and about events present and past, that Bruce can have no part in. They simply *appreciate* one another more than either of them appreciates anyone else, and they would rather be appreciated by one another more than by anyone else. They just are at home with one another, whether or not they can live together under the same roof, that is, find a roof they can live together under.

I mention several features of their intimacy which this film picks up quite unmodified from the laws of the genre of remarriage. There is the early, summary declaration that this woman has recently been created, and created by this man. What he created her from was a "doll-faced hick," which thus satisfies the law that they knew one another in childhood, anyway in a life before their shared adulthood. And what he created out of her was a newspaperman. This creation accordingly hinges with the further feature in which accepted differences between the genders are made into problems, several related ones. The conventional distribution of physical vanity, first of all, is reversed. Our opening glimpse of Walter is of him primping, and soon he will be giving himself a flower to wear, as though dressing for battle. It takes a while for Hildy's comparative casualness about her looks to reveal itself, an occasion, for example, to notice the way she grabs her hat and coat or peels them off for chasing a piece of news or for the work of hammering it out on the typewriter. The question which of them is the active and which the passive partner is treated at the close of their initial interview as a gag, as in *Bringing Up Baby*, about who is following whom, or about who should be. In *His Girl Friday* it takes the form of issues about who is to go first down the aisle through the city room and about who is to hold a door and a gate open for whom.

The gag is awarded one of the four beautiful long tracking shots the film allows itself, this one following the pair from Walter's office to the outer lobby to meet Bruce. This passage reverses in parallel the earlier passage that tracks Hildy's entering path from Bruce to Walter's office. But that earlier, preparatory passage had been even more explicitly rhetorical. The tracking motion was compounded there, heightened, by being handed back and forth, in a shot-reverse shot between Hildy's gaze and the successive gazes of those with whom she exchanges greetings. These exchanges affirm a special attentiveness the camera pays to Hildy, both in the lyrical impulse to follow her and return to her and in the knowledge it registers of the anxiety in walking back into the familiar room and greeting these inquiring faces after an absence of the four months in Reno and then Bermuda. But there is something more. The reverse shot of the faces Hildy greets is taken from a distinct point of view, I would like to say an implicitly occupied point of view, but it is emphatically not Hildy's. It is a point just ahead of Hildy's progress, as though something is anticipating her arrival, preparing the way for her.

This movement and position had itself been prepared by the preceding movement, the opening shot of the film, in which the camera moved the length of the city room from right to left, that is from what we will learn to be the direction of Walter's office. It stops at Rosalind Russell just entering, waits while she exchanges a few words with the telephone operators, a few more with Ralph Bellamy, and then accompanies her back to, we discover, Walter, as if it had come from him exactly for the purpose of guiding or taking her back. From this beginning pair of tracking shots, therefore, there is established the possibility that leading someone, as opposed to following someone, has its own gallantry.

I note that the title *His Girl Friday* implies a certain order or precedence between the members of the pair; it is evidently the case that, like the man Friday, Hildy is meant to do the work, or to be, as she puts it, Walter's errand boy. The implication that the pair are survivors, as in *Robinson Crusoe,* is perfectly apt to the tale we are to see unfold, as will emerge. But the title equally alludes, I cannot but think, to a popular radio serial of the period, "Our Gal Sunday." The daily narrative lead-in to each episode of the serial spoke of "a story about a girl from a small mining town in the West [a doll-faced hick?] which asks the question whether she can find happiness with a rich and titled Englishman." I do not require that serious viewers of this film accept a memory of this lead-in as a parodistic description of the relation in this film of Rosalind Russell to Cary Grant; any more than I require that such viewers recognize that in Walter's late line, "The last person that said that to me was Archie Leach a week before he cut his throat," Cary Grant is mentioning his original (English) name. Any good comedy is apt to like some inside jokes that it can afford to throw away. But the allusion to the soap opera title I regard as unmistakable, once it occurs to you, and I take it to signify that what we are to experience is an anti-soap opera, a work meant to challenge the words and moods of popular romance, hence to invoke those words and moods at every turn. (A relation is being established here to other Hollywood film genres.) An example of the working of the lingo of these films is Walter's impatiently muttering, "Oh, oh, now the *moon's* out!" as he is contemplating ways to remove the desk with Earl Williams in it out of the Press Room, thus invoking what Dexter in *The Philadelphia Story* will call a jealous goddess as in this context an inconvenient fact.

The restaurant sequence is the central time Hildy and Walter are seated together, conversing, rather than walking fast or talking fast. In the previous sequence, in his office, they momentarily perch themselves on the edges of things, such as desks, awkwardly, and the posture of sitting down is given a specific interpretation by Walter, as he holds his arms out to Hildy, pats his knees to indicate where he's inviting her to sit down and says, "You know there's always been a light burning in the window for you," to which she retorts, "I jumped out of that window a long time ago." As the film is closing in the Criminal Courts Building they are seated together again. But to say how this comes about, and what it means, we first have to know what place the Criminal Courts Building is, I mean what its place is in this narrative.

IT IS A PLACE seemingly so unlike any other place we witness in the remaining comedies of the genre of remarriage as itself to cast doubt on the placement of this film within the genre. In each of the other instances some pastoral alternative exists to the desperations of city life, whereas in *His Girl Friday* we move from the watchfulness of a city room to the sleeplessness of an all-night vigil in an even darker region of the city. This suggests a distinction between satiric and romantic comedy that would assign *His Girl Friday* to the satiric side, away from the six I wish to put it with. In Chapter 1 I had occasion to quote Frye's contrasting of Shakespearean and Jonsonian comedy to the effect that the latter inspired the realistic comedy of manners that became the dominant tradition of comedy on the English stage, and suggesting that Shakespearean romantic comedy may be said to have survived only in opera. Now listen to a different inflection of the distinction between Jonsonian and Shakespearean comedy as drawn by Nevill Coghill in

They simply appreciate one another more than either of them appreciates anyone else, and they would rather be appreciated by one another than by anyone else. They just are at home with one another, whether or not they can live together under the same roof.

"The Basis of Shakespearean Comedy." Jonsonian comedy is "The Satiric"; it "concerns a middle way of life, town dwellers, humble and private people. It pursues the principal characters with some bitterness for their vices and teaches what is useful in life and what is to be avoided. The Romantic [or Shakespearean comedy] expresses the idea that life is to be grasped [that is, not avoided]. It is the opposite of Tragedy in that the catastrophe solves all confusions and misunderstandings by some happy turn of events. It commonly included love-making and running off with girls"; "love is essentially an aristocratic experience; that is, an experience only possible to natures capable of refinement, be they high-born or low. In search of this refinement, Shakespeare began to imagine and explore what we have come to call his 'golden world' . . . It was a world of adventure and the countryside, where Jonson's was a world of exposure and the city." These are useful words. Allow me one more quotation: "Jonson's characters (representing ... humors ...) suffer no changes ... compared with the comedies of Shakespeare, those of Ben Jonson are no laughing matter. The population he chooses for his comedies in part accounts for this: it is a congeries of cits, parvenus, mountebanks, cozeners, dupes, braggarts, bullies, and bitches. No one loves anyone. If we are shown virtue in distress, it is the distress, not the virtue that matters. All this is done with an incredible, stupendous force of style."*

And yet for all the obvious and painful pertinence of these observations about the satiric to *His Girl Friday*, that film is still a story of the adventure of love. And while it is fully true of these characters that "they suffer no change," uniquely true among the members of the genre of remarriage, what we might call the pair's attitude toward this fundamental fact about themselves does undergo decisive alteration. It can be taken to be the goal of this narrative, which as elsewhere in the genre means the goal of the woman's education, to demonstrate that change in or by the object of her love is unthinkable, and that this is after all acceptable to her. The education will have its bitterness, but sweetness, apparently, enough.

It would be reasonable to describe *His Girl Friday* as the introduction of a Shakespearean leading pair into a Jonsonian environment. (This could be a way of describing what Hawks did to *The Front Page* in, as it is put, turning Hildy into a woman. The consequences he explores of this change are surely not less than those released in turning Plautus's identical male twins into one male and one female twin, as in the progress from *The Comedy of Errors* to *Twelfth Night*.) Such a description means little more than the justifications that might be given in its behalf, and I understand what I take to be the main justification for it to show the following asymmetry: the Jonsonian setting does not predict or require, if it is to contain a pair in love, that the love should be like *this* (so much might be taken as proven by *The Front Page*); whereas the Shakespearean pair does predict, or require, that if they are to inhabit a Jonsonian environment, it should function as the Press Room of the Criminal Courts Building functions. This, if true, allows me to keep and to use this film as a full member of the genre of remarriage.

The Criminal Courts Building occurs at a point, and plays the role,

* In *Shakespeare Criticism 1935–1960*, selected by A. Ridler (London: Oxford University Press, 1970).

that in at least four of the six other members of the genre is played by what Nevill Coghill refers to as "the golden world", what Frye calls, and I have been calling after him, "the green world". It is a place to which the action moves after an opening in a big city; the place within which the plot complicates and then resolves itself; a place beyond the normal world, where the normal laws of the world are interfered with; a place of perspective and education. But in this film this place is a terrible world, not golden, not green; a black world. The amusements in it are provided by joyless card-playing; by the voyeurism of Stairway Sam (the Shapely of *It Happened One Night*, still at it); by derisive wisecracks against both the defenseless and those in political power; by gallows humor. These diversions make for a certain camaraderie among the reporters, but the diversions of the prisoners elsewhere in the building would be composed of similar pleasures and make for an analogous camaraderie. Otherwise the setting is one of rumor, distortion, falsehood, corruption, brutality, a certain picture of the world of news, a certain picture of the world at large. The principal instrument of this world is the telephone, a forest of which dominate the table in the center of the Press Room. This instrument at once shows the reporters' attachment to this one thread of communication with the world outside, and their isolation from that world. This black world, finally, is presided over by a huge, central, artificial light, the inversion, or caricature, of the light of the natural world, the green one.

Notwithstanding, it is here that Hildy suffers her rebirth of feeling, prompted in particular by talking to Walter over their special distance of the telephone.

Go to the moment at which, furious with Walter for having had Bruce arrested on a phony charge, she is standing at the doorway, belongings in hand, and making her speech of farewell to the chumps of the newspaper game, delivering her declaration of freedom, of her escape to normal life. Machine gun shots ring out, and then a warning bell and a siren signal a prison break. These violent sounds of emergency are as if in response to Hildy's speech. They have a farcical, or symptomatic, aptness to Hildy's claim to be getting out, that is, breaking out. "What's going on?" a reporter yells out of the window. A voice from nowhere replies, "Earl Williams escaped!" Given a moment's thought we might almost laugh at the implied comment or conspiracy of the world, mobilizing itself to prevent Hildy from escaping, but there is no time for the laugh to express itself, or to recognize itself as such, so its energy further heightens the excitement of the moment. To the extent that our more settled idea is that the alarm is as of a conspiracy against her, or let us say, a piece of bad fortune, then the implied comment is that Hildy can no more escape this edifice, and what it means, than Earl Williams can. But to the extent that we read her as wanting the escape not from Walter but from Bruce, that is, an escape from her separation from Walter, then our idea of the alarm is as of a piece of good fortune, a perfect diversion to cover her getaway.

In either case, her fate is linked with Earl's, either to suffer it or to write about it, or somehow both. And in both cases Walter is behind all of this. We do not have to imagine that he could foretell or control specifically what would happen when he sent Hildy to this place, but he knows enough both about her and about this place to know that her fortune will strike home there. That is to say, he knows that her fate is to link up with Earl's.

The camera insists on this link from the early moments of her entry into the black world, her descent into the cell of death to interview Earl. Another pair of its few openly rhetorical moments are spent on this interview. The sequence opens with an extreme high-angle shot of her entrance into the space containing Earl's isolated cage; its last shot but one is a return to a medium shot of Hildy in profile, almost, as it is lit, in silhouette, immediately backed by the bars that Earl is immediately fronted by. The sequence reads to me, in outline, as follows.

The shots derive from German expressionist cinema, another homage from Hollywood. The point here is to declare the shots as meant to contain visual projections of a character's psychological state, and specifically as meant to turn the part of the world made visible into an experience of mood, a certain form of hauntedness. The high-angle entrance shot is, so conceived, inherently ambiguous. It is first of all a reflection of the woman's expansion of consciousness. She can see the whole of the situation, as we can see the whole of Earl's cage, including its top horizontal bars, emphasizing a prisoner's absolute loss of privacy, of subjection to visibility, as if to being filmed. The condemned man is not merely a story to Hildy; she is drawn to such a story. But at the same time the extreme angle expresses her distance, or estrangement, from her feeling, but in such a way as to prove that she is capable of genuine feeling, that this is what her struggle is about. The strategy of her interview is to manipulate Earl's untutored and tortured sensibility so that he produces a piece of nonsense she can use in her writing about him to dramatize his insanity and hence to make a case for granting him a reprieve. The piece of nonsense is to have him say that he fired the gun and killed a man because he had heard a soapbox orator speak of "production for use" and he had this gun and after all the use of a gun is to shoot. He is momentarily heartened by the sanity of the explanation. What her strategy comes to is to convert a socialist thought that is perhaps about a precommodity economy or perhaps about using the resources of industrial society to produce goods that meet real and comprehensible human needs into a thought of consuming a commodity whether one has a need for it or not, and in particular in such a way that it is destructive. Is the moral that socialist ideas are mad or is it that capitalist practice drives one mad? The reporters seem perfectly prepared to believe either, or rather both. Having victimized Earl, this time for his own good, and having gotten what she came for, Hildy withdraws into profile. She is turned equally from Earl and from us, private to the universe of news. She is alive to herself, to her reality, privacy intact, but not necessarily in possession of a name for what is going on in her.

She gets up, pauses on her way out, and delivers a remarkable line (I believe it was Manny Farber who first isolated it for attention): "Goodbye Earl. Good luck." I gloss the line, said then by this woman to this man there, in the following way: "I know you Earl and if you could know anything you would know me. We are both victims of a heartless world, and condemned to know it. The best the likes of you and me can hope for is a reprieve from it, on grounds of insanity. Good luck to us both."

This is my formulation of a piece of the knowledge of herself Walter dispatched her to acquire. But it is not quite hers yet; she still requires the experience of the forces arrayed against accepting the knowledge, the force of her struggle against giving up on what I called her counterfeit happiness, the panic in intending to divorce herself from her divorce of Walter. This is what she learns when the alarm is sounded. The others, the men, rush out, no more doubting the relevance of the emergency than prison guards would. She is alone, abandoned to her thoughts, captured by the moment; then she suddenly comes to life, starts throwing off her hat and coat, grabs the phone and asks for

Walter Burns quick, finishes getting stripped for action and, after filling him in, says excitedly, "Don't worry, I'm on the job!" We cut to a scene of her outside, overtaking Dooley, the death cell guard she had earlier bribed to get the interview with Earl, and stopping him with a flying tackle, making sure we know her to be a newspaper man. Then with a cut back to the still empty Press Room a new mood is set, a new departure prepared.

The abandoned telephones are ringing urgently, hungrily, and the camera moves solicitously, close-up (it is the fourth of the tracking shots we counted earlier) along the length of the table they line. The movement is from right to left, as in the camera's opening gesture of the film; and, as in that gesture, it is dispatched to await one, the same one, whom it will thereupon stay with. The film is as if making a new beginning. As the camera waits, having moved away from the door and now watching it from the opposite end of the table, Hildy is not the first to enter. One reporter, then another, arrives and departs having fed the telephones with "more slop from the hanging," tidbits side-lit from a story these reporters have yet to find. When Hildy shows up, closes the door behind her for privacy, gets Walter on the line, and says to him, "I've got the real story, I've got it exclusive, and it's a pip. The jail break of your dreams . . . ," we witness a new tone in her conversation with Walter, or doubtless an old one. She is cooing at him. The tone is one of intimacy, one that other people normally reserve for intimacies, or that normal people reserve for other intimacies. She is as if making a new beginning. ("The jail break of your dreams" indeed. There is no doubt in my mind that this refers both to Walter's wishes as an editor and to his desires as a suitor in having sent Hildy to find her fate, that is to find him, in this place.) And after Earl enters through the window, crazed, waving the gun, and fires in response to a snapped window shade, Hildy grabs the gun, and then grabs the phone, calling breathlessly into the line to her newspaper, to whoever is on the other end, "Tell Walter I need him." This completes her education. She knows where she is.

We have, as viewers, by now received instruction from the film about where we are as well, I mean about where it places us, which means, as it does in each of our other films, how it places its own images, accounts for their provenance and their presentation. We have, for example, been given to think about some relation the camera is proposing between itself and those otherwise abandoned ringing phones as it, along with them, awaits a human being to respond, to them and to it. Fiction-ally, of course, there is an invisible someone at the other end of the phones, but in fact it is we who are invisibly at the other end, calling upon these figures, called by them. This seems to me confirmed by the times we cut from Hildy's end of a conversation to insets, as it were, of Walter or of Bruce, to witness their end; each time it seems to me that we are as if displaced by their appearance, though of course only in order that we might be called upon in turn by them.

A key to this film's placement of its images, or displacement of us before them, is to understand the light it gives us to see by. And here we have to consider what that monstrous, sadistic examination light is that Dr. Egelhofer, the alienist, turns on Earl as he examines him. (If it is supposed to be taken as a light by which prisoners are ordinarily inter-rogated then the doctor's sadism is merely an extension of the ordinary sadism of that world.) This first of all raises the question not merely, as Hawks for one among countless many had asked before, whether the psychiatrist or his patient is the crazier, but whether he or this patient who has killed a man is the more violent. But Hawks implicates us in this question by making Dr. Egelhofer's hateful examination light, with its enormous black tub reflector, a duplicate of the ceiling lights that occur in each of the rooms of the Criminal Courts Building as their sole source of illumination. According to my memory, these lights occur in just about all the shots in the Press Room but I call special attention to a repeated camera set-up in which one ceiling fixture hangs, as it were, in the middle of the frame and from its upper edge. I infer that we are shown these events, from the moment the lights in that room are nota-bly turned on, under the same illumination Dr. Egelhofer goes by. Then we are to ask ourselves how our understanding of its subjects differs from this inquisitor's understanding, how our satisfactions in knowing them show a firmer sense of reciprocity with them. The light under which we examine these figures thus becomes a light under which we are subject to examination. We would do well, accordingly, to make this too an examination *by* ourselves. This strikes me as another parable of criticism, of its obligation to say something for the experience of-fered to it.

If we provisionally accept this definition of ourselves, ourselves as viewers, functioning within the lights of this film, in the position at once of inquisitors and of victims, as Hildy is, then can we understand ourselves as needing and hoping for, using the words I used to express Hildy's farewell to Earl, a reprieve from a heartless world? What, a little more spelled out, is the heartlessness of the black world of this film?

If we say that this specifically means that this world is not fashioned according to the heart's desire, and indeed that it cannot so much as project for itself a world so fashioned (although it does hope for reform; as it puts it, for a new government like LaGuardia's in New York), this just amounts to repeating that this is not a Shakespearean comic world of romance. It nevertheless has its pleasures. We alluded to two of them earlier in speaking of Walter's capacities, the search for truth and for adventure. Finding and publishing the truth is a source of pleasure even if you cannot make the truth prevail, and it is itself described and depicted as an adventure. That it *can* be an adventure means that the world is still, however corrupted, knowable, and the truth of it publishable, and hence that both the truth and the world have a chance.

A related, derived pleasure, and a further part of what allows there to be adventure, lies in knowing the real value of things. This knowledge, which I earlier called knowing the falseness of the face of the world, is epitomized in an exchange that strikes me as immensely funnier than it has, I believe, struck the audiences I have attended it with. Hildy has just phoned Walter telling him that she has the real story and says that it cost her 450 dollars which she wants back right away. Walter covers the mouthpiece of the phone and says to Louis, his crooked Puck, "I've got to have 450 dollars worth of counterfeit money," to which Louis immediately replies, feeling his breast pockets, "Can't carry that much Boss." Walter says, "No, no, I mean just 450 counterfeit," to which Louis immediately replies, "Oh sure, Boss, I got that much on me." What I think causes the particular hilarity I feel in this exchange is Louis's unearthly freedom from certain human concerns together with his easy, expert, if displaced assurance in his knowledge of the precise relation between appearance and reality. At the opposite end of the spiritual world there is Bruce's somewhat displaced, concerned praise of life insurance to Walter: "Of course it doesn't help you much while you're alive. It's afterwards that's important," upon which Walter laughs heartily and then stops abruptly saying "I don't get it." Now nobody could be more absolutely this-worldly than Walter, so it is hardly surprising that this representative of life as adventure finds the placing of greater importance on after-life than on aliveness to be incompre-

COUNTERFEITING HAPPINESS

The high-angle entrance shot is first of all a reflection of the woman's expansion of consciousness. She can see the whole of the situation, as we can see the whole of Earl's cage, including its top horizontal bars, emphasizing a prisoner's absolute loss of privacy, of subjection to visibility, as if to being filmed.

hensible. (I shall not forbear noting that an assertion of this-worldliness, that is of mortality, is one of the most insistent of Desirée's attributes in *Smiles of a Summer Night*, the burden, for example, of her beautiful song in the final act of the drama, in the pavillion. I do not wish to harp on her relation to Walter, but this conjunction of references to a secular world suggests a further line of investigation in thinking of these comedies, a suggestion of mankind abandoned to itself.)

I think, beyond this, that the idea of an "afterwards," as Walter must repond to this, means what we mean by "consequences," and it is the characteristic of what we will doubtless read as Walter's amorality that he seems to lack altogether the concept of his actions as having moral consequences. If one does not consign him to moral idiocy, then his transcendence of consequences makes him the embodiment of the idea of life as improvisation. And here again the concept of improvisation, which I keep finding fundamental to the experience of film, is fundamental to the understanding of what one of its significant instances is about.

This heartless yet not quite hopeless world also has ugliness in it, not just the ugliness of obvious brutes and bullies, but a subtler kind, more pervasive. An instance is Walter's appearance at the instant Molly Malloy throws herself out of the window and his apparent refusal to respond to this act. It will be hard at first to avoid interpreting this refusal as his sheer indifference to suffering, or a refusal to be distracted by anything that makes no difference to his immediate plans. This feels ugly, as if something ugly is being done to us in being shown such a moment. A moment that seems to me to bear comparison with this revelation is Falstaff's stabbing of Hotspur's dead body. For each of these problematic sources of fun, it is an absolute truth that when life is over everything is over. There is freedom in this view, but also cruelty; we are instructed not to romanticize it. I think, beyond such significance, this brutal, ugly conjunction of events is an acknowledgment from Howard Hawks about the nature of the control a film director exercises over the viewer of film, in particular of the way in which the process of editing, the power to conjoin in principle anything with anything, can at will grab attention by the feet and dash its brains out against a flickering wall.

Perhaps, on further reflection, Walter's apparent indifference is a refusal to jeopardize an urgent plan, itself a matter of life and death, by attending to something to which he can make no difference. The others who rush to the scene can do nothing for Molly that is not already being done by those already there, and Walter will not stoop to treating the merely painful as though it is important as news. Walter's world is one of present crisis and catastrophe, where such things happen. "This is war, Hildy," he will soon say to convince her she must stay with him on the job. A normal person will understand this to mean that at such a time normal concerns are in abeyance. But for Walter there is no other time. Does this mean that for him nothing is normal? Then it must equally mean that for him nothing is abnormal. Hildy would have understood this, in her way, when she said to him, "You wouldn't know a half-way normal life if you saw one."

Walter might even assume Molly knew what she was doing, even admire it as a successful piece of improvisation, not sympathizing with it exactly, but seeing it as an image of his own actions. The trickster knows that he is open to consequences, to their trickiness, if not by the tricks of others then by those of fortune; he merely cannot wait to act upon their calculation. When at the conclusion he puts himself in the hands of an unseen power he in effect declares that his own power is only mortal, without certainty, without insurance; and he had early on expressed his feeling that Hildy had tricked him into marrying her in the first place—not, evidently, that he minds one way or the other, as long as they are together, but facts are facts. And is it a fact that Molly's leap from the window is an improvised diversion? The act allows her to protect the other two victims she is allied with, to absorb their suffering into her own. "Why are you asking her where Earl is? She don't know, only I know . . . Now you'll never find out." It is with such words, screamed at the crowd of bullying men, that she leaps. So we know also that the gesture is one of rage at those men, and I take it as a gesture of ugliness, to curse them. This seems to me to establish a further bond between her and Walter, permitting his acceptance of her act. He need not feel that its accusation is leveled at him; he may even take confirmation from it. (In this connection remember Hildy's early statement to Walter about her having jumped out of a certain window a long time ago, roughly to save herself from him. This is a curious confirmation that Hildy's alignment with Earl is purified by Walter's, not her, alignment with Molly.)

Perhaps Walter's world of war is not exactly a normal world, but it is, as Thomas Hobbes and John Locke had painted it, a state of nature.

LOCKE, it is true, in his *Second Treatise of Government* objects to "some persons who have gone so far as to confuse the state of nature with a state of war" whereas they are as different from one another, in his words, as "a state of peace, goodwill, mutual assistance and preservation" is different from "a state of enmity, malice, violence and mutual destruction." This does not say that the state of nature *is* a state of peace; and later in the *Treatise* Locke describes the condition of the state of nature as, while free and equal, "full of fears and continual dangers," which is the reason, after all, men are willing to quit this condition and seek to join in society with others. But Locke's late description sounds suspiciously like Hobbes's understanding of the state of nature, which he identifies as a state of war. For Hobbes seems to mean by speaking in this regard of war something like fear and danger, or insecurity and uncertainty. He says: "as the nature of foul weather lieth not in a shower or two of rain, but in an inclination thereto of many days together; so the nature of war consisteth not in actual fighting, but in the known disposition thereto, during all the time there is no assurance to the contrary." We might try thinking of this as a lack of life insurance. "Whatsoever therefore is consequent to a time of war, where every man is enemy to every man; the same is consequent to the time, wherein men live without other security than what their own strength and their own invention shall furnish them withal." We might think of this invention as the power of improvisation. Still, there is good reason for Locke to object to the identification of the state of nature with a state of war. His aim is to understand the ending of a state of nature in its replacement by a civil state based on a contract of consent. This is not a conceivable ending of a war. Whatever one's conceptual doubts about this source of America's ideals, it is altogether important about Locke's notion of a state of nature that it was not merely or simply a useful conceptual fiction but equally a matter of experience. It is he, in the century before David Hume made the question more famous, who says, "It is often asked as a mighty objection, 'Where are or ever were there any men in such a state of nature?' "; to which he responds that "the world never was, nor ever will be, without numbers of men in that state." He instances Princes (that is, absolute monarchs) as being in a state of nature with respect to their subjects, because there is between them "no common, higher appeal"; and he instances on the same ground the relation of independent governments to one another.

Is there a "common, higher appeal" between the members cast together and made visible in *His Girl Friday*—between Earl or Molly and the reporters or the sheriff or the mayor, or between any of these and Walter? These are not safely academic questions for, let us say, Americans (of whatever origin), whose creation and continuation, invoking questions of union and secession, was and is always in doubt, as befits a noble experiment. Then the concept of marriage, understood as remarriage, as a search for reaffirmation, is not merely an analogy of the social bond, or a comment upon it, but it is a further instance of experimentation in consent and reciprocity. Thus does marriage again become of national importance.

Walter's late appeal to an unseen power is ambiguous as between an invocation of the "common, higher appeal" which is the mark of the existence of a state of society and that "appeal to heaven" which Locke takes as the mark of a state of war, for example, of the call to revolution. Walter speaks to Hildy as though revolution is at stake; and when he sketches a vision for her of her chance to throw out a corrupt government and put in a new; and describes this as "moving up to a new class," a remark which elicits from her a mysterious and startled, "Huh?"; it is quite as if Walter is promising that she may be the mother of her country, or anyway be something that is just as good, if not the same, as being married. But of course Walter's claim that this is war has no privileged knowledge as to whether Hobbes's or Locke's fantasies better fit our society. Indeed it is deliberately ambiguous in its fictional position. What is the "this" that is war? Does he mean reforming the state; or getting a story; or moving into a new class; or is it merely love, in which, as in war, all is reputedly fair?

I MENTIONED in my reading of *The Lady Eve* that *His Girl Friday*, notably among the comedies of remarriage, does not end, even by implication, with a request for forgiveness by the man and woman of one another, and I suggested that this is a way of understanding the terrible darkness of this comedy. It is a darkness shared explicitly in our genre, and there not so relentlessly, only by *It Happened One Night*, which is also the only other member of the genre without a localized green world. But there is, in each of the others, some glimpse of an ugliness in the world outside, within which, or surrounded by which, the actions we witness take place. In particular, a glimpse of the failure of civilization to, let me say, make human beings civil. Each shows a world in which beings view others as objects of entertainment and scandal, as unequal to themselves, and would exclude others from civilization, treating them with a civilized ugliness (the father's treatment of the mother in *The Philadelphia Story*, Sidney Kidd's treatment of anyone and everyone, the mother-in-law and the fiancée's parents in *The Awful Truth*); or they show beings whose weirdness suggests that civilization has been unable to recruit them as equal to itself, who cannot be imagined to survive outside the particular environment that knows them (the Sheriff and the Colonel in *Bringing Up Baby*); or they show worlds in which lawlessness and order as a whole are explicitly in struggle (*The Lady Eve*; *Adam's Rib*). Then the issue of these comedies is how we are given to understand the relation of the pair we predominantly care about to this surrounding world, how it is that they escape its evils sufficiently to find happiness in it.

Walter Burns's answer, as said, is the capacity for adventure, so far construed as the capacity for improvisation, for the lack of insurance. (Improvisation and the capacity for taking risks are characterized as virtues in an early essay of mine entitled "Music Discomposed.")* But this is at most a necessary, not a sufficient, condition for his happiness, which requires the presence of Hildy to share the adventure. And the world must have room in it for this capacity. For what is adventure without a world?

While this black world is without forgiveness, it has in it, as we know, and as required by the law of the genre, its equivalent, namely the possibility of reprieve, a real, if in each case temporary, relief from the pain of the world. And the temporary might be as good as the permanent if it lasts long enough or recurs reliably enough. How do we get a reprieve? Where does it come from?

It comes via Mr. Pettibone, who comes, he says, from the Governor. It should give us to think how it is that this apparently witless, hapless being can have found his way to the Criminal Courts Building, and even somehow been in touch with the Governor, who he seems the only one to know has gone from fishing to duck hunting, since the Mayor as well as the *Morning Post* has been trying to get the Governor on the telephone for days. When Mr. Pettibone returns at the last possible moment and makes himself clear at least to Walter, Walter's elated response is to cry, "What did I tell you? There is an unseen power that protects the *Post.*" These are elaborate hints that we are to give the character of Mr. Pettibone and his errand sufficiently mythological attention. Dropping in at just the last moment with the item to

* In *Must We Mean What We Say?*

prevent catastrophe, this spirit comes not from a machine exactly (unless you are willing already to accept the presence of the film image, and its power of credibility, as machine enough), but from a doorway through which he is trying to maneuver an open umbrella or say parachute, or say caduceus, as of Hermes, the messenger. His touch of madness and good humor is still intact, as if these qualities are the surest protection in the world from the world's corruption. He seems to me the spirit of comedy itself, responding to our insatiable desire for happiness, acceptable in the form of reprieve. As for the Governor, or unseen power, from whom he comes, I appeal to the characteristic acknowledgment the directors of these films give of their presence in them by use of the concepts of magic and of invisibility.

Evidently some relation is being proposed between Bruce and Mr. Pettibone. Both carry umbrellas through thick and thin, both are family men, and both pride themselves on being honorable. Mr. Pettibone reappears only when Bruce has gone for good. Is this to remind us that the values Bruce represents are, after all, indispensable to our salvation? Or is it to confirm that comedic luck, to which happiness is tied, cannot appear as long as we try to tie up fate with insurance? Bruce carries his umbrella for a prudential reason; if it doesn't rain the thing is useless, an appendage, an excrescence. Mr. Pettibone carries his for spiritual reasons, out of his sense of his identity, to ornament his worth, as in an excess of energy. Mr. Pettibone's relation to Bruce associates him with Hildy, thus suggesting that it is Hildy's presence that brings Walter luck.

But how are we, who are we, to require and to receive the benefit of this reprieve? I have so far only tried to indicate why it is that we, who are condemned, as I imagine Hildy to be acknowledging in her salute to Earl, to know the heartlessness of the world, stand in need of reprieve, that is, how it is that our position as inquisitors makes us victims of our own viewing. But actually to receive the benefit of the reprieve we have to develop what it means to interpret our position in being viewers as a position of being victims.

Victimization constitutes an interpretation of the passiveness of viewing. In "On Makavejev On Bergman" I propose that recent films of Bergman and of Makavejev show how the action of what is exhibited by projecting images on a screen may be treated mythologically as what Nietzsche calls, in *The Gay Science*, action at a distance, which is his interpretation of a man's reaction to women (or, as I suppose, of the relation of the masculine to the feminine side of human character). This proposes a mythology, in a word, of the seductiveness of film; of art, therefore, to the extent that film is art. Bergman and Makavejev understand the screen, according to this way of thinking, as reflecting for us a way of considering things that presented in themselves would turn us to stone. Call these things elements in our horror of our sexuality, of our existence with others in the world. The point of the myth is that our condition as passive, as victim, might damn or might save us, might darken or illuminate us, depending upon whether we are impassive or receptive to the experience offered us, closed or open to it. I want the idea of receptiveness here to hark back to the mark that Heidegger, and I have claimed Emerson before him, requires of genuine thinking.* Naturally I would like to understand what I mean by reading a film as a mode of this thinking.

I have proposed that what constitutes a reprieve from the pain of the world is what Walter Burns means by the capacity for adventure. Then Hildy's reprieve is her acceptance of the adventure with Walter, of the old adventure, for we have already noted that no change is in view. But what is this adventure for her? Is it merely, for example, as the concluding image suggests, their running out of the door of the Press Room, him first, to cover a strike (in Albany, too!) instead of finally going on a honeymoon? She is holding the bag, a fact emphasized by Walter's saying to her over his shoulder, "Shouldn't you carry that in your hand?" This situation is no doubt a little insane, enough to serve as grounds for a reprieve. But let us consider what an improvised honeymoon of this kind means to them. We know from their recent, relieved exchange about a former tight spot (in which they could have gone to jail for stealing a stomach [it was a piece of evidence in a trial] and in which while they were unmarried and hiding out for a week they stole pleasures for which, as Walter reminds her, they could also have gone to jail), as well as from their reminiscences in front of Bruce at the restaurant, that their adventures covering stories on the road have been the occasions of memorable sexual encounters between them. In such a case who, one might think, needs what the world calls a honeymoon?

And then there is, at almost the last words of the film, Walter's at last

* "Thinking of Emerson," in *The Senses of Walden*, expanded ed.

sitting down to continue his frenetic talk with Duffy about preparing the front page for the story Hildy is hammering out at his side. He sits down at the moment and as the expression of his conviction that Bruce is leaving without Hildy. Their being seated together at the same table means what it meant from the beginning, that they are at home. And if they can be at home in that black world, they are at home anywhere, at home in the world. They have achieved the goal of romance. (It is this fact of their being home, rather than their plain isolation from the rest of the world, an isolation that is emphasized in all of the other comedies of remarriage save, for good reason, *The Philadelphia Story*, that claims their relation to Robinson Crusoe and Friday.)

So she has a home after all, the one that adventure can give her. It is not all of happiness, but nothing on earth is. And what happiness there is has demanded, as suggested, her conspiracy in the ruthlessness of its pursuit. The most telling conspiracy is her acceptance of Louis's packing off over his shoulder the woman she calls Mother. This is the literal, brutal form in this world in which a mother disappears in order that a marriage may happily supplant itself.

We do not really know that the pair are going off together unmarried; probably they do not, as we see them leave, know either. The speculation is pertinent. It is a premiss of farce that marriage kills romance. It is a project of the genre of remarriage to refuse to draw a conclusion from this premiss but rather to turn the tables on farce, to turn marriage itself into romance, into adventure, which for Walter and Hildy means to preserve within it something of the illicit, to find as it were a moral equivalent of the immoral.

"A reprieve from the world on grounds of insanity." This now sounds to me like a characterization of comedy. As though we recognize in the insanity of comic events an image of the insanity of our lives in requiring such a reprieve. A colloquial version of the idea of comedy as reprieve is the idea of film as providing "escape," which seems to be the most common public understanding of what film is good for, though I have not heard it said what the escape is from, nor where it is to, nor what its method is. In a film such as *His Girl Friday*, one which is all but an allegory of what film is, one which studies a director as a passive trickster at the other end of a line, an actor as a victim of visibility, an audience of viewers as victims of their passive knowledge of these things; a film which is a member of a genre of film that turns out to demand of each of its members its own interpretation or study of these matters; we are asked to consider further what we shall say the effect of film is, the good of it, by considering what specific films of significance show it to be. This is an obligation of what I mean by reading a film. But this consideration will be formed by how we appropriate the work of film, say actively or passively; in the present instance, by how we take its idea of reprieve. In one way it may be taken as escape (in which case you must keep on escaping); in another way it may be taken as refreshment and recreation (in which case you are free to stop and think).

THE
COURTING
OF
MARRIAGE

Adam's Rib

What does he imagine a Punch and Judy show is, and why does he imagine his wife might turn a courtroom over to one? The film's daring, complex capping of this question depends on seeing the resemblance between the curtains of a puppet stage and the curtains of a four-poster bed.

A S if reversing the condition of the world of the pair in *His Girl Friday*, the pair in *Adam's Rib* (1949) are emphatically at home at home. George Cukor thoroughly details their inhabitation of their private world for us—we are invited into every room in their two-story apartment, from living room and study and kitchen to bedroom and dressing room and bathroom, and we witness every interaction between them from sexual invitations and drinking and cooking together to massaging one another. Our presence there becomes so natural, returning each night, that we recall with surprise that this is, with one minor exception, the only member of our genre in which we see the pair in their own home at all. The exception is *The Awful Truth*; it is a minor, if significant, exception since in it the man and woman enter the house separately, each with other company; and upon finding themselves alone they talk a little of what Cary Grant calls—and what any sensible person would call—philosophy; whereupon they decide they must divorce.

In "More of *The World Viewed*" I note the careful establishing and rewarding of our intimacy with the central marriage of *Adam's Rib*, singling out particularly the sequence in which the camera is fixed (as we in our places) on the pair's bedroom, now empty, save intermittently, of their presences, and they speak to one another from their opposite dressing rooms, hers just left and out of the visual frame, his just right. "The effect is to increase our intimacy with these figures because their invisible and pervasive presence to us puts us in the same rela-

191

tion to each of them in this passage as they bear to one another."* Here
is this film's allegory of the nature of viewing film. The sense of participation or partnership in their intimacy is essential to the way the film
works, because it is exactly this intimacy that the woman puts on trial
in taking her marriage to court. We will not understand her bravery
(nor, hence, the man's) unless we know that for her their intimacy, their
privacy, their home at home, is almost everything. Not to call it or to try
to make it everything is doubtless something that makes it so good;
then faithfulness to it requires that it be capable of being held at risk.

But why now? We know Amanda Bonner (Katharine Hepburn) takes
the defense of a case because her husband, Adam (Spencer Tracy), as
Assistant District Attorney, has been given the case to prosecute. But as
she says, in expressing her decision to her secretary, the fact of her
husband's having the case, or accepting the case, is "the last straw on a
female camel's back"; so the burden had been piling up. What is the
nature of the burden? And along what axis has whatever it is been piling up? Is the last straw some further misdeed of her husband's, in participating in a society's systematic wrongs? Or is it some further misdeed of society's, in requiring her husband's systematic participation?
Somehow both, since, as a successful American lawyer, she is hardly
unaware of the mutual implication in one another of the life of society
and the life of the law, and hardly, in general, disapproving of that mutual implication. But their mutual implication means that one is no better than the other, and the suggestion is clear enough that the institution
of marriage can be no better either, that it is part of that implication.
The question is whether, and why, any of them is good enough, inhab-
itable, bearable on a female camel's back. A condition of their bearable
imperfection is that marriage *can* be taken to court, that it is *subject* to
debate.

In describing what Amanda does in challenging her husband on the
case of Mrs. Attinger (Judy Holliday)—whom we have seen, in a kind
of prologue to the film, shoot her husband (Tom Ewell), having followed him to the apartment of another woman (Jean Hagen)—as taking
her marriage to court, I am assuming from the outset that the legalities
of the case remain obscure throughout the film, strung, one might say,
between Amanda's impulse to excuse Mrs. Attinger and her impulse to

---

* *The World Viewed*, enlarged ed., p. 200.

justify her. I am assuming further that Amanda is not merely bringing charges against her own marriage but simultaneously questioning whether courts, anyway as they stand, are capable of assessing the validity of marriage. Adam will, in the lecture to his wife that our genre makes obligatory, begin with the question "What is marriage?" and answer, "It's a contract, it's the law," and go on to imply that her disrespect for law will end by leaving nothing to respect. But he is, as she has determined, really sore at her and says some things that his anger wants to hear, such as that he is not so sure he any longer likes being married to the so-called New Woman, and that he is old-fashioned enough to like there to be two sexes. These are, from him and to her, about as dirty as remarks can get. And quite incoherent. How is she denying the legality of marriage? Of the Attinger marriage she will claim the next day (as if to parody Adam's lecture) that it is sufficiently sacred to justify, to protect itself by, assault with a deadly weapon. Or is Adam suggesting that she is denying or betraying the legality of her and Adam's marriage? He knows her well enough to know that all the protection his marriage needs, from inside, is an assault with a licorice pistol.

(The incoherence upon Adam's characterization of marriage as a contract underscores explicitly, in this latest of the definitive remarriage comedies, that there is no such genre apart from two social facts: that divorce is regarded as possible, a morally and religiously acceptable option; and that we remain unsettled, accordingly, about what makes marriage an honorable estate. The genre of remarriage may be said to find the humor in this state of affairs preferable to the humor derivable from a state of affairs in which divorce is not a moral or religious option, to the farce, say, in adultery. The word "contract," at this climactic moment, to my ear names the social contract that was to express the consent that constitutes lawful society, the doctrine that replaces the divine right of kings. Here again the fate of the marriage bond in our genre is meant to epitomize the fate of the democratic social bond, as more or less explicitly in the aristocratic equations of marriage and society in *The Philadelphia Story*, or the equation of victim and wife and heroine in *His Girl Friday*, the linking of fates that underlies, as I argued, Milton's argument for divorce. I rephrase these matters parenthetically here because the recurrent doubt has struck me again here, as it might at any time, whether we fully recognize how remarkable the problematic of these comedies is. It is not remarkable to be told publicly that the integrity of society depends upon the integrity of the family. But it is something else to be told that the integrity of society is a function of the integrity of marriage, and vice versa, where marriage is validated neither by a family nor by the law.)

I understand Adam to feel (not exactly that the legality of his marriage has been infringed but) that a bargain, let me say, of their marriage has been broken—something like a bargain that his wife will not oppose him publicly, professionally. Such an issue between them is almost the last matter they discuss as he tells her she wouldn't run against him for a judgeship because he'd cry and she would have to respect his tears, even fake tears ("the old juice" he had called them when they were hers). I suppose that his side of this bargain is his support, publicly and privately, in an imperfect world, of her feminist convictions. Then not only must she feel that his accepting of the prosecution of Mrs. Attinger is an original breaking of their bargain, she also feels, anyway after the fact, that the bargain had itself been a mounting burden, itself awaiting a last straw.

Because she is not treated as a doll at home, she has been willing to conclude that she has not been living in a doll house. But the absolute division between home and world establishes a late version of Nora's sense of confinement. What Amanda wants of the world that as a professional woman she does not already have is evidently for the world to know that she is an equal at home, an equal in intimacy and in authority. And evidently she wants this knowledge because she wants not only something more in the world but something more at home. She wants her husband's knowledge of her acceptance outside, of her public separateness. Independence inside and outside reverse one another, hence require one another.

So she brings inside and outside together, her marriage and the world, in the space of a courtroom. Her husband says, in trying to dissuade her from taking the case, that he's not going to let her turn a courtroom into a Punch and Judy show. What does he imagine a Punch and Judy show is, and why does he imagine his wife might turn a courtroom over to one?

This is a question for us as much as for him since the film *Adam's Rib*, which presents the tribulations and trials of a marriage as a source of popular entertainment, including the pair's slugging and kicking one another, itself takes on the color of a Punch and Judy show. That this is part of the film's self-understanding seems to me declared in its animated titles, and in its inter-titles (used mostly to inform us that scenes to follow take place later that evening), whose background is a curtained stage, I imagine a puppet stage. The film's daring, complex capping of this idea depends on seeing the resemblance between the curtains of a puppet stage and the curtains of a four-poster bed, in front of which the concluding moments of the film are played and into which the pair will disappear, pulling the curtains closed behind them. The suggestion is that the marriage bed is the final setting of a Punch and Judy show, and I take the image to stand for everything in marriage that is invisible to outsiders, which is essentially everything, or everything essential. In this most elaborate revelation of the life of a pair at home, in which we felt privileged to be behind the scenes each night, apparently sharing the residues of one day and preparing the terms of another, we wind up with a curtain drawn in our face.

If Adam means something vaguer or more colloquial by accusing Amanda of looking to turn the courtroom into a Punch and Judy show, say that she wants to make some kind of mockery of the sanctity of the law, what is the justice of his charge? She is aggrieved and he doesn't see why; his not seeing why magnifies the grief, is part of the grief. The reciprocity of marriage makes it a fertile field for revenge, understood as getting even or as teaching a lesson. But instead of taking private revenge (as perhaps Doris Attinger was doing) Amanda Bonner turns to the law; in this she is acting in the true spirit of the law. She still wants a lesson to be taught, something she calls "dramatizing an injustice." She has to mean this as a justification for risking Doris Attinger's future, using her problem as an occasion for opening a public issue to a public verdict. She is equally using that problem, I suppose with less awareness, as an occasion for opening a private issue, hers with her husband, to a private verdict. As said, this ambiguity ensures that her legal argument will remain obscure. The public demonstration, for society's instruction or self-confession, is a reasonably clear feminist manifesto to the effect that women are the equal of men in intelligence, in accomplishment, in responsibilities, and hence deserve equal rights. Her private demonstration, for her husband's edification, can only be determined by her behavior in the courtroom.

Before turning to that behavior let us again ask why it is now that she joins the issue with her husband in this way, which is to ask why it is now that she breaks what I called the bargain not to oppose him publicly. This way of putting the question rules out the simple answer that an opportunity has presented itself; that answer simply denies that a faithful bargain had been in effect. I take it that she has discovered that he *wants* this case, and wants it for private reasons, hence as a demonstration to her. An explicit part of her evidence would be his remark over the telephone, having called to tell her that he has been assigned the Attinger case, "You're cute when you get causy"—an attempt at levity which succeeds merely in being dully dismissive. He realizes this, but he doesn't want to see that this time he is part of the cause.

And I understand his wanting of the case to be confirmed by her implicit knowledge of him, the mode of wordless knowledge intimates harbor of one another, intangible maybe but as consequential as bad moods. As she wakes him, in our opening view of them, having received their breakfast tray, with the morning newspapers on it, from the housekeeper at the threshold of their bedroom, she tells him he had been making sounds in his sleep; she imitates them for him, objectively, as somewhere between sounds of desire and of pain. She takes it that there is something on his mind, in his dreams. And the images here suggest to me that she understands what he has been moaning and groaning about in his sleep to be what is recorded in the newspaper story she reads to him. He has said, "You always say that I always do" (make noises in the night), and she replies, "You always do, but...," and then instead of describing how this time it was different, as a proper narrative would, her eye hits the newspaper as with a force of revelation. (It turns out that the story is not in his newspaper; anyway not up front.) A newspaper flung before their door, picked up and carried with the rest of the ready morning comforts upstairs to the locked door of even deeper privacy, exposes the Bonner constitution to the Attinger discomfiture.

Something is being laid at their doorstep, something from the depths, social and psychic, the stuff dreams and responsibilities and entertainments are made of. Adam and Amanda argue about the significance of the newspaper material from the moment, from before the moment, they get out of bed; their instincts are elicited on opposite sides of some line. Then when Adam is assigned responsibility for the case it is as if Amanda attributes to him a dream of the case, an unacknowledged instinct or brief against women as the source of what he will call, in his cross-examination of Mrs. Attinger that Amanda will interrupt, "your shrewishness, your domestic failure." He makes what I take to be an equivalent charge against Amanda late in his late lecture to her about marriage, as he is packing to leave, when he asks her what people watching them think of them, and answers, "They think we're uncivilized. Uncivilized." (Freud accuses women of threatening civilization because of their centripetal interest in their own family. Adam's accusation, on the contrary, seems caused by Amanda's centrifugal interest in civilization.) Adam participates in the dream of the male world, and to this extent shares his instincts, however refined his expression of them, with the likes of Mr. Attinger.

To contest Adam's dream of women, hence of her, Amanda has to confront her marriage and its world with one another, to let them rebuke one another (like America and American law). This requires that her marriage and its world each be good enough and sound enough to profit from this exposure; their acceptance of the exposure will be the best proof of their value. The husband's exposure will require, as in the genre of remarriage it must, that he undergo a certain humiliation, a dunking of his dignity (the air going out of Gable's tire; Grant's being dressed in a negligée or covered with feathers; Fonda's repeated fallings; Tracy's dizziness and stuttering in court). His capacity to permit himself to seem ridiculous, without thereby losing his sense of worth, is what makes him worth listening to, what gives him the authority to lecture the woman, to be chosen by her for her instruction. Call it his ability to learn, to suffer change. In showing that he allows, and survives, the going out of his ego, this ability proves his potency.

I am trying to find a cause of Amanda's actions that I can believe in. To believe in her immediate partisan excitement ("instinctive" is the word she will use about Adam's male brutality) upon reading that a woman shot her husband ("Kill him?" Adam inquires; "Nope. Condition critical, though."), I might alternatively imagine that she regards the situation of American women after World War II to be equivalent, morally and psychologically if not materially, to, say, that of southern slaves before the Civil War, where one slave will readily be imagined to have such an instinctive partisan reaction to news that a fellow slave had snapped and revolted against yet another outrage. A sympathetic outsider might well have an analogous reaction. Or I might try imagining the case on analogy with that of a woman whose country is under foreign occupation, or a military tyranny, and whose husband is beguiled by enemy glamour. The fact is that I do not believe this is the way things are for American women, or were soon after World War II, and I cannot imagine that Amanda Bonner believes it either. I do not at all mean that the situation of women in America around 1950 or around 1980 cannot from time to time *strike* one as similarly outrageous, or insupportable. But this perception, which I take to be valid, differs in two respects from the outrage of those suffering the oppressions of slavery or of foreign or domestic occupation: first, the perception is not stable, but comes and goes, which means that the situation is not simple but mixed; second, the stimulus for an occasion of being struck by the outrageousness of the situation may be something comparatively trivial, perhaps a rude remark, or perhaps its dawning on you, for no apparent reason, that your culture assumes that doctors and lawyers are men while nurses and secretaries are women—something that reveals a process of arbitrariness and injustice. Mr. Attinger—or of course Mrs. Attinger—may have had good cause for straying from home; but nothing is good cause for injustice.

But why *at all* try to get Amanda's instinctive reaction so that it is realistically believable? Why not just take it as what certain critics call a "premiss" that she did have the reaction, and let the story take it from there? But what good is a premiss if it is not believable? I am merely working out the consequences of accepting it. I think that an impatient sense that I am being too literal-minded or reality-intoxicated here would amount to a sense that I ought not to make certain sorts of demands on what is after all only a movie. But whatever the merits, or the meaning, of such a sense, it is irrelevant to the point I have just now been addressing, which is about my *interest* in a passage of this movie, about my experience of it, about the part of my life I have spent with it. To consider what it would mean, what it would look like, to defend the taking of an interest in one's experience, perhaps the best thing one can (still) call one's own, was a guiding task of my Introduction. The task deserves all the attention it can get, an essential part of which, for me, is to let it question my progress whenever it must—question whether indeed the progress of my prose is everywhere faithful to its implicit claim to be checking its experience, monitoring its economy, a term of criticism I accept as pertinent to my ambitions.

THE FORM the pair s contention takes in the courtroom is one that pits the woman's attempt to make something public against the man's attempt to keep something private. The cause of the contention, the thing she wants, is his awareness of her, his recognition, let me say his acknowledgment of her. While her whole point in being in court is to make something private public, he repeatedly appeals away from the public to their accustomed privacy.

He is still at it on the last of the four days of the trial, as he calls her Pinkie in open court, leading to some confusion on the part of the court reporter. She knows, he knows, we know at this point, that she has won, that he is not able to fight her as an equal in public; and we know that to demonstrate this is at once the form and the content of the trial, her vindication, at once her public and her private victory. But he had made his most memorable and charming effort in the direction back from the public to their privacy early the first day in court, as he silently invites her at the opposite end of the lawyer's table to do as he does and accidentally on purpose knock a pencil to the floor so that they can have a moment to exchange wicked glances of appreciation under the table. The natural, even the logical, enmity between the erotic and the legal, noted differently when Walter Burns reminded Hildy of what they could have gone to jail for, is the explicit plot of George Stevens's *Talk of the Town,* in which the two paths of feeling and of law are presented (by Cary Grant and Ronald Coleman) as equally attractive, equally noble (if not equally respectable), but as mutually exclusive, between which the woman (Jean Arthur) has to choose. Just to make things inescapably clear, the man of feeling is portrayed as an anarchist. The comedy of the romance of remarriage is the idea that such a choice may not have to be final.

An emblem of Amanda's response in kind to Adam's courtroom conduct, or her dramatization of the thing his conduct is a response to (her appeal away from their privacy to the realm of the public, the direction explicitly renounced and explicitly longed for by Tracy Lord), is her costuming of Mrs. Attinger in the hat Adam had given her before

At least two doubts immediately present themselves in the face of the attempt to think of the depicted home movie as a melodrama, first whether there is a narrative here at all, second, if there is, whether it contains a villain, a despoiler of virtue, without which there can hardly exist a melodrama, and hardly a virtue, worth the name.

he knew they would be in court together. It is, one might say, her main exhibit on her side of the case, apart from Adam himself. It was a genuine present but also a real enough bribe, buying her silence toward his work of prosecution. She exhibits the hat, accordingly, as a rebuke to the bribe but also because she is proud of her husband's way (as opposed, for example, to Mr. Attinger's way) of expressing himself to her. So that when, in Adam's stammering summation, he tears the hat from Mrs. Attinger's head and pockets it, entering as an exhibit the receipt that shows the hat was bought by him, he is exactly confirming

Amanda's charge of him, both in and against his favor. She does not conceal the tremendous pleasure she takes in this demonstration.

THE BATTLE OF THE SEXES, in its form of a struggle for recognition, especially for the woman's recognition, is the guide to the cinematic presentation of the courtroom sequences, particularly of the second day in court, the longest, in which each of the pair of lawyers examines, successively, Beryl Kane, Mr. Attinger, and Mrs. Attinger.

The battle for awareness is pervasively depicted in eye-movements, especially Hepburn's. In Amanda's examination of Doris Attinger there is the following set of events. Amanda says to her, of her discovery of her husband and Beryl together, "It enraged you," upon which Adam interjects from behind her, "Objection. Leading," whereupon we watch Hepburn's eyes dart left, then right, as if glancing over each of her shoulders to inspect Adam. Then she walks as it were directly away from him and without looking at anybody in particular, asks, "When you saw them thus embraced, what happened?" but the direction of the remark, so to speak, has just been established by her eyes. The multi-directionality of her courtroom communication is as if diagrammed for our instruction when, in the opening shot of her examination of Mr. Attinger, she is facing the jury, her hands on the railing of the jury box; addressing a question to the witness to her left; but carrying on a conversation with her husband behind her. It is her show; her future happiness depends on eliciting the right responses from each of these audiences.

Adam's reciprocal task is sketched in such a set of events as the following, still within Amanda's examination of Doris. On Doris's close-up testimony that as she kicked open the door what she saw was her husband "Nuzzling that tall job," she throws a glance out toward the courtroom, as if pointing briefly with her chin. It is a conventional sign that we are now to expect either a point of view shot of that tall job or else a matching shot of her reaction to that epithet. Instead Cukor holds the camera a moment longer on Doris, then cuts to an over-all shot (anyway clearly not a point of view shot) of the whole courtroom, with no one in particular singled out or even clearly locatable. Then after again this shot is held longer than our expectations would predict, Adam rises to object. I read this pair of automatisms as follows. The brief fermata on the close-up of Doris's face is long enough to call up an anxiety about what is to happen next; this anxiety is increased by the succeeding fermata over the whole courtroom. Then when Adam rises to object we find words for our anxiety. Adam was failing for a couple of difficult moments to pick up his cue. Or better, we understand that he is losing his sense of when it falls to him to say something, of what constitutes a cue for him. His disorientation generally, of course, is that one part of him insists that the case is one of something like attempted homicide while the part of him that is bound to his wife's part insists differently. More specifically, so far as he is called upon to answer her demonstration to him, he is at a loss for words since, as she said to him the previous night, she knows he agrees with her convictions on something to do with these matters. So what does she want him to say? Nothing special, or nothing in particular. Knowing quite well what it is that he thinks, what is on her mind is rather, as in bringing them to court in the first place, *whether* he will say anything at all, whether there is anything about this situation that he can publicly claim conviction in. If not, that is her victory, whatever the depicted jury will say.

CUKOR'S CAMERA INSTINCT, unlike Hawks's, is to move in response to a character's attention. In *Adam's Rib* we have in effect noticed two variations on this relation. First, in the shot of the mostly vacant bedroom, flanked by their conversation, the camera, I would like to say, *abides* in response to its equal attention to each of them. Second, we just saw the camera take a position on a character's behalf not because it has been called upon by his attention but as if it is calling his attention to the fact that is is up to him to say something, to come to. I note that in the sequence just discussed Adam rises to speak from behind Amanda, as it were over her shoulder, the direction in which she has been speaking to him. One might consider whether this should be read further as his arising *from* her shoulder, as she from his rib.

Assigning significance to the stations and the progressions of the camera, and to whether its movement in a given case is small or large, toward or away, up or down or around, fast or slow, continuous or discontinuous, is something I mean by reading a film. It requires acts of criticism that determine why the cinematic event is what it is *here*, at this moment in this film; that determine, indeed, what the cinematic

event *is*. A camera cannot in general *just* abide or progress, *just* be continuous or discontinuous; it has to abide on something, and move or
dissolve from something to something. As the mind cannot in general
just think or the eye just see, but has to think *of* something, look *at* or *for*
or *away from* something. Phenomenologists speak of the mind, in that it
takes objects, as intentional. I should like to speak of the camera, in that
it takes subjects, as inflectional. (But really speaking in abbreviation of
the entire chain of wish and apparatus that leads from director through
camera to projected image as the work of a camera is already speaking
of film as inflectional, or say as photographic. A filmmaker who wishes
to defeat this inflectionality will wish to defeat the camera, or his dependence on the camera, in arriving at his projected images.) Reading a
film may accordingly be said to require understanding the motivation
of the camera, accounting for its inflections, its modifications. Why it
modifies itself as it does on a given occasion, why it bends and how it
warps when and as it does, what it is responding to inside or outside
itself, it is the business of film criticism to determine. To say this about
film criticism is in turn a remark of film theory, whose business more
generally might be described as specifying the existence of the camera,
its possibilities and necessities, how it can be motivated in the ways
criticism determines it to be. Theory may, among other pieces of business, have to provide a characterization of "motivation" here that will
justify or replace its anthropomorphism. None of this says how we are
to grasp the relation between film criticism and theory, or shows that
they are separate studies.

I do not want to go again into the consequences of such powers of the
camera for the concept of reality, reality as the address of the photographic, which is the principal subject of "More of *The World Viewed*".
In the reading of *It Happened One Night* I characterized these consequences as the transcendental conditions of viewing film. And I know
that some who think about film think that from a recognition of the
powers of the camera to modify—from what I am calling the camera's
inflectionality—it follows that the camera does not present us with reality. I have heard offered as proof enough of this obviousness the thrill-
ing refrain, "Things as they are are changed upon the blue guitar," as
though something is obvious in these words, as though their point and
beauty were something beyond the poise of their ambiguity. What are
changed are exactly things as they are, perhaps as from one hand to

another, or as from a strum to an ear. The changes upon a guitar are its progressions, its harmonic motions, changes upon itself, which, Stevens's words thus claim, still take as their object things as they are, which must therefore be such as to lend themselves to these changes, be changed by them, as by a serenade. This is not an affirmation of, say, Kant's ceding of things-in-themselves, but a contesting of that gesture. The last thing this poetry takes for granted is the nature of itself as an instrument, the nature of its changes; the last thing a camera, or an ambitious student of the camera, should take for granted is the nature of the camera. Perhaps it is, or in certain hands is, more like a pitchy viola, or a Beckmesserian lute. (J. L. Austin was thinking, like Wallace Stevens's line, about the internality of words and world to one another when he asked, parenthetically in his essay "Truth," "Do we focus the image or the battleship?")

I WAS LED to these speculations, breaking off my determinations of the camera's inflections or allegiances in the courtroom sequence—an investigation that my various examples of it are meant to suggest ought to be completed for each of our films, shot by shot—by a thought concerning the most extended sequence in *Adam's Rib* in which the camera declares itself, the showing of the home movie. In the course of depicting the projection of the home movie, its projection all but gets identified with the projection of our movie (the one entitled *Adam's Rib*), its frame made almost to coincide with our frame. To me this conveyed the thought that the study of the camera cannot be exhausted by the determination, however complete, of what I was calling the motivation of the camera, that is by film criticism (so conceived), but must invoke the question of the existence of the camera as such, that is by the issue of what I called film theory. The reasoning or provenance behind this thought is as follows. The near coincidence of the two movies implies that the role of the camera in the one is fundamentally no different from its role in the other. But since the motivation of the camera's progression in the home movie is without interest, or rather shows this critical question to reduce to the theoretical issue of accounting for the camera's existence, accounting for its presence altogether, the theoretical dimension must in general, in films like *Adam's Rib*, remain an issue after criticism has said what it can. (To escape criticism by invoking directly the theory of an art can be taken as a motive of the modern in art.)*

The home movie in *Adam's Rib* is entitled *The Mortgage the Merrier: A Too Real Epic*. Let us sort out some similarities and then some differences between *Adam's Rib* and *The Mortgage the Merrier*.

First, they share the same principal actors and characters and certain minor characters as well. Second, they both employ inter-titles in their narrative continuity. Third, while the contained film exhibits "Censored" as an inter-title and the containing film (minus the contained film) does not, each exhibits essentially the same incident of censoring, namely the man's suddenly grabbing the woman to maneuver her behind a door, out of our sight. Fourth, the question of the identity and role of the director is raised simultaneously for each film by the other just by their being in this container-contained relation to one another. The cultural invisibility of most Hollywood directors is being challenged by this Hollywood director in the act of framing the projection of a home movie in which a director is not only apparently invisible but apparently nonexistent. That these films ask investigation of one another is declared out loud when Kip (David Wayne), in his running commentary on *The Mortgage the Merrier*, asks, "Who took these pictures, your cow?" It requires an act of will not to take the reference of

---

* Since it is my view that, in what I call reading (a film), criticism and theory will eventually call upon one another, it should not be surprising that in this book, primarily devoted to criticism, theoretical questions keep presenting themselves, as in *The World Viewed* and its supplementary essay, theoretical claims are always meant to be substantiated by acts of criticism. I put these matters in a way I and some others have found usable in "What Becomes of Things on Film?" Because not everyone will have ready access to the journal in which it appears (*Philosophy and Literature*, published by the University of Michigan, Dearborn), I should like to reproduce its final paragraph here: "The question what becomes of objects when they are filmed and screened—like the question what becomes of particular people, and specific locales, and subjects and motifs when they are filmed by individual makers of film—has only one source of data for its answer, namely the appearance and significance of just those objects and people that are in fact to be found in the succession of films, or passages of films, that matter to us. To express their appearances, and define those significances, and articulate the nature of this mattering, are acts that help to constitute what we might call film criticism. [You cannot know the significances a priori, for example by consulting some code; interpretation is required.] Then to explain how these appearances, significances, and matterings—these specific events of photogenesis—are made possible by the general photogenesis of film altogether, by the fact, as I more or less put it in *The World Viewed*, that objects on film are always already displaced, *trouvé* (i.e., that we as viewers are always already displaced before them), would be an undertaking of what we might call film theory."

that question to be *Adam's Rib* as a whole, all of the "pictures" it gives us to see. Fifth, while *The Mortgage the Merrier* refers to a real event, part at least of what is filmed in it is a performance, I mean something that it itself recognizes as a performance. As one of the guests at the screening explains, who also happens to be in that movie, "We acted all this out later of course. It's not actual" (like that line itself? like the screening? etcetera). Sixth, the plots end in the same way, anyway in the same place, at the pair's place in Connecticut, and to that extent have the same setting.

The profound difference between the containing and the contained movie may be said to be that relation of containment itself. But is it more profound than the similarities between them? The French title I have heard for what I have described as containment is *mise en abîme*, placement in abyss; I think of this as endless displacement, a good phrase for the endless mutual reflections these films create for one another. I know of two examples given to illustrate the French words. One example is that of the facing mirrors in old barber shops, each of which reflects the thing the other reflects, and reflects the reflections of the other; and since, so to speak, one of those reflections is itself, each reflects itself, as far as the eye and its light can reach. The phenomenon is fascinating and the analogy is striking but, I find, misleading, since the point of the mirror-phenomenon is that the containing-contained relation cannot apply. The mirrors are equally originals. The other illustrative example of placement in abyss is that of a container which presents a representation of itself, hence a representation of something presenting a representation of itself (the Morton's Salt box). This much you can see; then the mind is drawn in after itself. Here the container-contained relation is preserved: the real box must be bigger than the biggest representation it contains. But this way of preserving the asymmetry of the relation is essentially inapplicable to the container-contained relation of the films, since the smaller, contained film is not a representation of the containing film. It, to go no further, contains no film. And both films are equally real, equally films; they have, so to speak, the same dimensionality.

Nevertheless the title *mise en abîme* seems to suggest to some theorists that we know of some adequate explanation or theory of this displacement. My guiding assumption is that everything we know of it must be derived from its function in particular films.

The containing-contained relation makes the containing film essentially more complex than the film it contains (since it contains something the other cannnot contain), but we do not know what significance attaches to this complexity. It may go with this to say that the home movie is meant for a private audience and the commercial movie for a public, but then to consider this relation is something we know to be a goal of the more complex movie as a whole. And again we do not know in advance of that goal what significance extends from this difference to, let us say, the nature of these movies as movies.

Then putting aside the greater complexity or elaboration of character, plot, and setting in the containing film, and the differences of audience each film expects, the remaining differences between them seem· to come down to two: *Adam's Rib* has better production values than *The Mortgage the Merrier* (or, more strictly, better values than the nonexistent home movie which the elaborately produced fictional home movie we are shown impersonates); and *Adam's Rib* is a talkie whereas *The Mortgage the Merrier* is not.

Finding the similarities between the movies to be no less significant than their differences, I wish to understand *Adam's Rib* to be acknowledging the autonomy of the film it incorporates, declaring it to be a complete (primitive) film on its own. In thus identifying its own enterprise with that of *The Mortgage the Merrier*, *Adam's Rib* is claiming the continuity of Hollywood sound comedy with two primary sources of the art of film: with the fact and the tradition of documentary film (of which home movies form a massive if peculiar species, and for which the home movie we are shown here suggests in turn a fictional basis), and with the fact and the tradition of silent film, especially melodrama (where the ownership and security of a mortgage can set the terms of plot, character, and setting).

I am prepared to draw the moral, anticipated a while ago, that no event within a film is as significant as the event of film itself. Significant how? No automatism, let me say, is as "cinematic" as the automatism of film as such. Since this moral can depend upon nothing beyond accepting the revelations of a film such as *Adam's Rib* as significant revelations about film as such, it is pointless, or rather too pointed, to treat the moral as a thesis, as though something beyond a continuing allegiance to one's own experience and a continuing assessment of one's commitment to a canon of films can yield a credible, or rational, conclusion.

This still seems to leave open the option of denying the moral I just drew, the option of affirming instead something to the effect that the event or fact of film itself is not what is fundamental to the cinematic, but rather it is what is *done* with the facts of film that is fundamental. However, the apparent denial of my moral should be understood to confirm it, or clarify it. For what it means to say that the event of film itself is the fundamental cinematic event is that what the maker of film does with the facts of film (call this his or her style) is to reveal that event, to participate in discovering its unfolding significance, which only the entire history of an art could complete.

But if I am right that the acceptance of such a view rests on experiences typified by the mutual assessment of *Adam's Rib* and *The Mortgage the Merrier*, it will be clear that we will not always maintain a very strong conviction in this view, or this moral (those of us subject to these convictions), indeed that our interest in the state of this conviction will itself wax and wane. I find that I am at the moment interested to pursue that mutual assessment far enough to see its unendingness.

The *Mortgage the Merrier* is a document of marriage and of ownership, as other home movies will be documents of weddings and departures or of birthday parties or of other arrivals. An implication is that *Adam's Rib* is a companion document of whatever its subject will be found to be, say it is a document of remarriage, and a further implication is that a document of marriage will take the form of a document of ownership, or of inhabitation (in capitalist culture?). The reason for this conjunction must be, it seems to me, to remind ourselves that we think of marriage, or have thought of it, as the entering simultaneously into a new public and a new private connection, the creation at once of new spaces of communality and of exclusiveness, of a new outside and inside to a life, spaces expressible by the private ownership of a house, literally an apartment, a place that is part of and apart within a larger habitation. And here again, as always before, an explicit economic issue poses itself ambiguously or inconclusively. You are free to interpret the issue as showing that only those who have money enough to afford a private dwelling, indeed two private dwellings, are in a position to pursue marital happiness. You are also free to understand the economic issue as part, hence as trope, of a more general issue of human happiness, call it the task or the cost of joint inhabitation, an essential requirement of which is the mutual creation of room, the resources for which (economic, spiritual, epistemological, metaphysical, geographical) remain incompletely charted.

How could one expect, in a film about marriage, anything more definite here? No doubt one should not claim to be certain that ownership of private property is necessary to the possession of whatever privacy is necessary to human happiness. But I am prepared to say that to the extent that we are able to let this be an empirical question, human history has not shown that private ownership (of *something*) is not necessary. One might well say that something standing in the way of human happiness is a false privacy, or a false idea of privacy. But then that is a reasonable formulation of what I have taken the argument of *Adam's Rib* to be about, the perception that leads Amanda Bonner to take her happy marriage to court.

We could also formulate what she does as risking the turning of her romance into melodrama. That the melodrama of the Attinger prologue, its gun and its tears, is meant as a threat to the Bonner story is shown by the fake gun and the fake tears which come out to real effect in the Bonner story. And its threat is prefigured by what I described as the melodramatic topic of the home movie.

At least two doubts immediately present themselves in the face of the attempt to think of *The Mortgage the Merrier* as melodrama, first whether there is a narrative here at all, second, if there is, whether it contains a villain, a despoiler of virtue, without which there can hardly exist a melodrama, and hardly a virtue, worth the name.

As to the second doubt, it is sufficiently dissolved for me in remembering Adam's twice adopting the classical or cliché look of a villain, the first time as he is about to burn the paid-off mortgage on a barbecue stand and holds a hotdog under his nose as a villainous moustache; the second time as he is about to follow Amanda behind the door of their barn and he stops to point lasciviously in her direction, in a silent, conspiratorial aside to the camera, that is, to its audience. But can we seriously understand Adam's mugging and horsing around for a home movie to suggest that he is somehow a real source of villainy, a melodramatic threat to the romance of his and Amanda's marriage?

I understand this to be, though not exactly assertible, roughly the brief Amanda has against Adam. And while not comprehensible as a charge in a court of law, it is the charge she suggests against him at home, again twice. The second time as farce, as they give one another massages. Adam's turn at massaging concludes with her declaration that she can tell the difference between a slap and a slug, and that it felt as if he felt he had a right to hit her. He has behaved like—well not a villain exactly, but like a bully, or a brute, someone equally unworthy of the favors of intimacy. The sequence ends with her suggestion, "Let's all be manly," and kicking him, thus equating manliness with brutishness hence being a man with being a villain. The first time with mutual seriousness, as she and Adam are making dinner and he asks her to give up the Attinger case. She says she wants to dramatize an injustice, like the Boston Tea Party, and reminds him that they couldn't be so close if he didn't agree with everything she wants and hopes and believes in, or unless she believed he did. Otherwise, the implication is, she would feel violated by their intimacies, he would be a cad to accept them; he might as well be demanding them of her in return for paying the rent.

To control the idea that mugging and horsing around neutralize whatever offense the man may give, that they make his actions lighthearted gestures that anyone should be able to take who can take a joke, focus on the fact that he is mugging and horsing around for the benefit of a camera. His asides conspire with the camera and its audience, conspire with, in a word, society. And this is the classical position of the charming, expansive villain (Iago, Edmund). The declared presence of the camera demonstrates that the villain is a role sought by the camera, one of its natural foods, hence that the villain is himself victimized by the camera's appetite, as his victim is by his appetite, itself a further, an opposite range of the camera's. The conspiracy between camera and society, serving one another's desires, has taken effect before the villain joins it. (I do not wish, in trying for a moment to resist, or scrutinize, the power of Spencer Tracy's playfulness, to deny that I sometimes feel Katharine Hepburn to lack a certain humor about herself, to count the till a little too often. But then I think of how often I have cast the world I want to live in as one in which my capacities for playfulness and for seriousness are not used against one another, so against me. I am the lady they always want to saw in half.)

As to the first doubt in the face of thinking of the contained movie as melodrama, hence as threatening the containing romance with melodrama, the doubt whether it exhibits a narrative progression, this depends on a working understanding of narrative. Suppose one thinks of it as a discourse in which something happens, in which there is an event that makes a difference, and so entails a before and demands an after, or say entails a comparatively uneventful context of beginning and of ending. The form of narrative Kip's running commentary takes is that of the travelogue ("Barn-kissing, an old Connecticut custom"; "and as the something sails off into the sinking something we say goodbye to . . ."), which allows a minimal difference between the context and the event that stands out from it, since the context is exotic, hence from the beginning has the interest of an event. Evidently the most important event of the home movie is the very screening of it. The screening commences with the crisis of Adam's spilling the tray of drinks upon hearing Amanda say to one of the judge guests that she is going to defend Doris Attinger, and it concludes with a cut, or really with the more intimate conjunction of a dissolve, to Adam and Amanda's bedroom and her shouting voice, "All right, all right, all right! You've said the same thing nine times!" So the mild fun of the home movie occupies the place of the event of difference that starts a narrative, flanked by emotions of crisis and spanned by Kip's grating narration.

What, more concretely, constitutes this event? There is, immediately, the sheer self-reference of the contraption of movie projection itself, generally by the very fact of depicting the home movie; more specifically, having all but identified its frame with ours, by containing certain notable breakdowns to which the contraption is subject, for example, sprocket misalignment, and a sequence misspliced upside down.

Of the unending set of questions about the medium of film such self-references may call to attention I single out here the question as to the nature of the surface of the medium. A writer about film who thinks to do justice to the fact that film is (presumably among other things) a visual medium may try to think of film as if it were essentially painting. One form of this attempt is or was to take certain concepts as deployed in certain "formalist" art criticism of the fifties and sixties as though terms were simultaneously being defined for photography and film as well as for painting (a procedure the art critics I have learned most from must deplore, since they were as much at pains to distinguish the conditions of the various arts as the various arts themselves were).* Such writers about film would sometimes speak of something called "the

* By the art critics I have learned most from I mean primarily Michael Fried and, through him, Clement Greenberg.

surface of the screen." It is worth saying that this phrase has no clear meaning and then going on to say, what the home movie of *Adam's Rib* serves to declare, that a screen *has* no surface but *is* a surface. Any surface that can hold the light of projection is (can serve as) a film screen. This fact may or may not have ontological significance, depending on how one makes out ontological significance, but it is what makes possible the special closeness and special distance between a film's depicting a projection of a film (the frame of the depicted film is then less than the frame of the current projection), and a film's, so to speak, projecting a projection (where the frames coincide), which just comes to projecting that film. (The photograph of a photograph, their edges coinciding, is the same photograph, in a later generation, it is a duplicate. A painting of a painting is not the same painting; it is a good or bad copy. As you can learn something about painting by working alongside someone who can paint, so you can learn something about taking photographs by working alongside someone taking photographs that you care about. But while for a significant history of painting you could learn painting in part by copying paintings, at no time in the history of photography could you have learned photography in part by photographing photographs, if this is different from printing them.)

The shift from depicting to projecting a film-within-the-film is a clear acknowledgment of the fact and the nature of film (for example, that a film is something made for projection and that not every way of showing it counts as projecting it), as clear and significant in *Adam's Rib* as a similar pairing of automatisms in Vertov's *Man with a Movie Camera*. But accuracy here is of the essence, and we can be more accurate. The sequence of the film-within-the-film in *Adam's Rib* is organized by alternating shots showing the home movie screen with shots showing the home movie audience, never mixing these subjects. The first two times the home screen is shown its screen is distinctly smaller than ours; while our shot contains no other equally distinct subjects, the top border can be seen to be occupied by a stretch of ceiling and the top segment of a column which are part of the room in which the screening is taking place, visible beyond the depicted screen. In subsequent shots of this screen we have moved closer, nothing whatever is visible beyond the depicted screen, but yet the frame of that movie is not allowed quite to coincide with ours. So it is not quite accurate to say about the film-within-the-film of *Adam's Rib*—something it would be accurate to say about a film-within-the-film in *Man with a Movie Camera*—that we are given a shift from the depicting of a film to the projecting of that very film by identifying ours with it. Rather, in *Adam's Rib* we shift from a clear case of depiction to a position in which it is ambiguous whether we are meant to understand the film as depicting or as projecting the film it contains. This is as explicit an acknowledgment of the medium of film as the related shift in *Man with a Movie Camera*, but what it acknowledges, when it is expressed, may be, however close in manifest technique, unpredictably distant in latent content. In *Adam's Rib* the acknowledgment of the nature of movies is a route for acknowledging the reality of its actors, declaring the people in the home movie, Katharine Hepburn and Spencer Tracy, to be the same as the actors in *Adam's Rib* who are watching themselves, and who are playing the parts of people watching themselves, in a home movie.

BUT WE SHOULD GO BACK to that other realm of circumstances that constitutes the event of the screening of the home movie, Kip's running narration of it.

His narration casts him as a film critic, hence affirms that what he is criticizing is a film. George Cukor and his script writers Garson Kanin and Ruth Gordon are having fun here with two kinds of critics. Directly, by showing on Tracy's glowering face how richly the author of bright, compulsive wisecracks about the sentimental vulnerabilities of one's life deserves a slap in the mouth. Less directly, by producing the sort of pedant, or village explainer, whose remarks cry out for the wisecracks—the sort who says, "Of course we acted all this out afterwards. It isn't actual," to which Kip rudely but satisfyingly replies, "All right big mouth, settle down." This is a reply, in turn, made for certain of today's pedants, big-time explainers, who offer to save us from falling for things like film's illusion of reality (as if, for example, the phrase "illusion of reality" is somehow clearer than the phrase "real reality").

It is to the point that this wisecracking critic is also portrayed generally as a moocher, mooching off other people's parties as well as off their privacies. (I don't know what to make of the fact that Kip is supposed to have written a bad Cole Porter song, beyond observing that mooching normally demands some talent and a certain bankable charm. But had a good Cole Porter song been attributed to him I would have a harder time characterizing him in the terms I do.) I draw the moral that a critic of such films as *Adam's Rib,* as of such as *The Mortgage the Merrier,* is bound to be a critic of marriage, of marriage as projected and criticized by the films themselves.

And this must work to create in Adam's eyes the connection between Kip and Amanda. We cannot imagine Adam to take Kip's attraction to Amanda seriously; he must understand that Kip's is an attachment, not uncommon around certain strong, interesting marriages, to the marriage itself, not to the woman alone. Nevertheless Adam barges in upon them together. I think the reason is that Kip's mild show-biz homosexual tinge is meant to align with Amanda's underlying charge against Adam's brutishness or caddishness, the charge of villainy by virtue of being a man. (This alignment is broadly hinted at a couple of times. After singing his song for them Kip says to Amanda, for Adam's benefit: "You've got me so convinced I may go out and become a woman." This remark also serves to prepare later speculations about sex reversals, which we will come to in a moment. Again, during Kip's narration of the home movie, as Adam is captured kneeling winningly between their two bulldogs, and as someone in the company exclaims something helpful like, "Oh, look. Dogs," Kip slowly counts them: "One, two, three.") As if to combat the unanswerable brief of being a villain because a man, Adam portrays a male bully, threatens double murder, and then victoriously mocks their portrait of him, and by implication their threat to him, by eating his gun. Is this savoring his revenge, said to be sweet; or is it successfully swallowing his anger along with his pride?

LET US NOT UNDERESTIMATE the depth of myth touched upon in these events, in their efforts to sketch the difference between men and women. Adam attributes his success in getting what he wants, getting Amanda back, to his ability to produce tears at will, that it, to a certain theatrical talent. But we have also seen his theatrical talent exercised in his entering menacingly with a (stage) gun, which also got him what he wanted, Amanda's acknowledgment that, as he says to her, "No matter what you think you think, you think the same as I think, that I have no right, that nobody has the right to break the law." And this meeting of minds more earnestly constitutes, for both of them, his getting her

Hatted, as for departure, they resume their adventure of desire, their pursuit of happiness. This pair is inventing gallantry between one another.

back. Are we certain, then, that his brutishness has played no essential role in his triumph? Both his gun and his tears play the role, demanded by male competition in the films of our genre, of the man's explicitly making a claim upon the woman. (And in what genre is this moment not demanded of a romantic hero? In such a work as *To Have and Have Not*, where it is the man who has to become reborn into the world in order—still—to prove or provide innocence, Bacall says to Bogart, "I'm hard to get, Steve. All you have to do is whistle." And she is right, and sympathetic. It is hard. In Hawks's earlier *Only Angels Have Wings*, Jean Arthur uses virtually the identical words to Cary Grant; and she is right, too.)

The man's tears play to the woman's maternal or tenderer instincts, the gun to her tougher, even vulgar, demand to be won. The point of the gesture of the gun is that intimacy is not sufficient for marriage, which requires beyond this the open declaration of this exclusive privacy. Openness is required as a condition both of asking for the public sanction of the marriage, admitting society's stake in it; and of expressing the need for this stake, that their bonding requires a decision, or contract, and power to have it enforced, that it is not natural, not, so to speak, a family matter. The simultaneous establishing and transcending of intimacy, the translation of intimacy, say as from strum to ear, is a way of putting the point of *Adam's Rib* generally, its interpretation of the dialectic of remarriage, why it is good to think of the necessity of remarriage as the necessity of taking marriage to court: you must test it in the open or else mutual independence is threatened, the capacity to notice one another, to remember beginning, to remember that you are strangers; but it is only worth subjecting to this examination if the case is one of intimacy, which you might describe as the threat of mutual independence.

After the trial, as Adam and Amanda are walking together out of the courtroom, Amanda says she wishes it could have been a tie. And in a sense it has been a tie; she has won the day, made her private point public, but he has won the night, made his public point private. It is what they each most wanted of the world, of one another. But the resolutions, the victories, are not stable, the tie is not yet, or not again, that of marriage. Or say that the separate victories are too stable. The conversation has not resumed.

When it does resume it will wind up on the topic of the difference between and the sameness of men and women. We have seen the conversation come to a halt in several stages, principally in the sequence in which Amanda comes home late, with a present of her own, to a silent Adam and an empty cocktail shaker. She follows him through each room of their apartment pleading with him to talk to her; and when he does open his mouth it is to deliver himself of his longest speech, the haranguing aria about marriage as a contract. The use of words in this pair of incidents is capped the next night as Adam's demonstration and victory with the licorice gun becomes a mostly verbal brawl. An angel would have difficulty at this moment distinguishing their lives from the lives of the Attingers; which is roughly to say that for an angel there is no distinction between comedy and tragedy. Their subjection to human commonness holds an important piece of learning for the Bonners— that while civilization has more to go on in their fortunate lives than in the Attingers, civilization cannot carry its own guarantee, either to last or to be worth the name.

That there is no humanly envisionable conclusion to the conversation of marriage seems to me the message of the mysterious moment at which, as Amanda in her summation is asking the jury to apply the same unwritten law to a woman as it would to a man, she directs them to imagine the principals of the trial triangle as reversed in sex. As she urges the jury to focus their attention on each of the three faces in turn and to concentrate, as if hypnotizing the jury, the faces in turn alter before our eyes. The immediate effect of the process is to seal once for all our identification, as audience, as a jury, for whose instruction and judgment the entire film is put forth. In taking the transformation effort as evidence that no conclusion to the problem of marriage is envisioned, I mean to imply that the transformations are not successful, anyway that they do not do what Amanda wants them to do, namely help us imagine these figures with their genders reversed.

In fact the transformations are, I find, grotesque, partly no doubt because the two women are transformed into pretty young boys and the man into something older, harder, coarser. Here is a reminder that the playing of one another by the sexes is not fully symmetrical. Boys can play women (as in Elizabethan theater) and women boys (as in opera), but for a mature male to impersonate a female requires the mastery of a significant art; and I know of no male impersonators, none I mean who are female. (The background distinctions here are among male and female impressionists, reactive heterosexual men, and what you might call imitation or fake men. This last is a quality that people, doubtless mostly certain men, may see in Hepburn, people who do not recognize what she is and the dimension a certain boyishness gives to her way of being womanly. Much of this is duly noted in Doris's remark during her interview, in response to Amanda's offer of a cigarette, that she doesn't think women should smoke, that it isn't feminine, that Amanda should excuse her for saying so, and our having to go ahead and imagine Hepburn/Amanda reacting to the source of the deprived observation, and allowing herself to excuse it.)

The asymmetries here suggest that Amanda cannot get what she wants from her experiment, not at any rate by the means she uses. The camera's transformation of the sexes of the characters seems to me a violence done to them, and to us in witnessing it, and to Amanda in asking it. She has, as Adam accuses her of doing, staged and costumed Doris's performance in the courtroom. But she has not presided over her transformation by the camera. She has only invoked a power beyond her control, and the results seem to me to show that she is being rebuked for it. Rather than casting her as a surrogate for a film director, the actual film director is depicting limits to the powers of an actor.

I HAVE SPOKEN of Cukor's willingness to let his camera follow the attention of a character, like a good listener, shunning imposition here. Let me call such followings the changes of progression. What I might call the changes of transformation, the translation at once, as a whole, from flesh and blood into film, is something else. It is the source of the camera's original violence, hence of the film director's original responsibilities. Cukor had studied this explicitly and at full length in *A Woman's Face*, of 1941, the year after *The Philadelphia Story*. (And again with Donald Ogden Stewart as writer, joined by Elliot Paul for *A Woman's Face*, perhaps because of the European setting.) The transformation of actress into star, as a version of the theme of the alteration of imagination in *The Philadelphia Story*, was something I touched upon principally respecting the relation of actor to audience. The complement or supplement, respecting the relation of director to actor, is the transformation of which *A Woman's Face* can be taken as a parable. It is again a story of a woman (Joan Crawford) drawn between two men (Conrad Veidt and Melvyn Douglas), that is to say between her love for two men. The choice in the romantic melodrama of *A Woman's Face* seems easier (granted a happy ending) than in the romantic comedy of *The Philadelphia Story* since in the melodrama the other man is openly presented as a villain. But this should be understood as a piece of cinematic code for a kind of love, against which it is by no means easy to choose. (One of the tinniest of a small chain of tinny moments in *A Woman's Face* is the woman's declaring in court that her feeling for the other man was not love: "I know that now," says Joan Crawford.) And again the woman is to be transformed, created, by the man, born into the world, but this time not from above the world but from below it, not from cold beauty but from seething disfigurement. The film studies the opposed processes by which the transformations are to be effected; it is this that makes it a parable of directing. (I daresay it is this sort of study that critics have responded to in calling Cukor a woman's director.)

The villainous process is portrayed as something like hypnosis, or enchantment, worked by a man of cultivated words, of demanding eyes and hands, and of a mastery of moods, invoked by his virtuoso piano playing. The heroic process is portrayed as plastic surgery, its completeness of reconstruction underscored by the surgeon's repeatedly calling the woman his Galatea. Cukor's acceptance of the metaphor of plastic surgery for the work of his camera seems to me the meaning of the set of moments in which the camera prevents us from seeing what the results of the surgery have been by shooting at an angle that leaves an obstacle exactly between us and the area of the disfiguring scar, and then, when the results are to be revealed to the court by her removing her hat, and after repeated set-ups in which the doctor examines her face under the merciless light that we see to be the light by which she is there being subjected to the camera, we are shown her face in unobstructed close-up, in one of those romantically lit portraits, drawings in light and shadow, that Hollywood was so good at in its high period of black-and-white, glorying in the assured results of its own work. In contrast to the heroic process of cinema, the villainous feels very like the process of theater.

The evil of the villainous procedure is that while it promises the woman release it leaves her unchanged, above all sealed in the isolation of her moral disfigurement, appealing to the realm of the demonic and its vengefulness which she has learned to call home. The good of the heroic procedure is that the point of the excruciating physical pain is to leave the matter of spiritual change up to her; the doctor repeatedly asks her whether he has created a monster or a woman, appealing to the realm of her better angel. This is why his direction is therapeutic. The removal of the scar redeems the marked woman, potentially makes her whole again, innocent. The creation of innocence through the right forgoing of virginity—such is the fantasy around which more than our genre alone has formed itself. (It is not surprising to find it in Garson Kanin's *Born Yesterday*, filmed by Cukor in 1950; an advanced New Comedy. And I hear *film noir* declaring its allegiance to the fantasy when in the closing moment of Robert Siodmak's *The Killers*, a bad, hysterical Ava Gardner kneels over the unconscious body of Albert Dekker, screaming, as if to call the words back from the grave, "Say

Kitty is innocent. Say Kitty is innocent.") I note, for further geographical reference, that the roles of hero and of villain in *A Woman's Face* are given respectively to an American and a European type. Melvyn Douglas is sophisticated, and in the fiction of the film is said to be Swiss, but set next to Conrad Veidt and Albert Basserman he is as American as Berlin, Connecticut. War was here, or imminent, and Veidt says something about how he will use the fortune Joan Crawford is to help him inherit in the service of the new order in Europe, but this is a topical cover for a preoccupation in American fiction given its highest form by Henry James.

Pat and Mike (1952) is a gentle, summery anthology of these themes of Cukor's as well as of other possibilities we have found in remarriage comedies. Like *Born Yesterday* it is what I called an advanced New Comedy, by which I mean essentially two things. First, its structure is to get the woman away from a false or outworn authority (the senex) by the help of a man who wants her to take authority over herself. That what is wrong with the senex is not a matter of his literal age is emphasized by Spencer Tracy's being called "old man" by the one the woman calls her "beau." Second, the film displaces virginity well lost with, say, integrity, or selfhood, as the goal of the drama, a new beginning. *Pat and Mike* makes this displacement possible by having Hepburn call herself a "widow"—about which Mike's sidekick (Sammy White) understandably is puzzled: "Ain't that like not married?" It places these issues in a context in which an all but literal directoral function is given to each of the two men. Her beau is anxious about how she will act or behave or perform and whether she is properly attired; Tracy calls himself her "manager and promoter," gives her principles for pacing herself, tries to remove obstacles to her coming through as herself, and reassures her with a director's privileged words: "You're a beautiful thing to see—in action." And the film fuses with these, further, the feature of the man coming from a lower class than the woman (always likely to be an issue for what becomes of Katharine Hepburn on film), winning her giving of herself to him by performing his remarkable feat of awakening her or freeing her. The issue of the creation of the star and the woman is dwelt upon lovingly, repeatedly through a fairy tale set of three questions. Mike catechizes his other star, a prizefighter (the affecting Aldo Ray): "Who made ya'?" "You did, Mike"; "Who owns the biggest piece of ya'?" "You do, Mike"; "What'll happen if I let go of ya'?" "I'll go down the drain"; "And?" "Never come back." At the close, Pat turns the tables by asking the questions of Mike. He answers the three all right—thus acknowledging that the star she is is the cause of his being the manager he is, as well as vice versa—but then turns the tables back with the capper: "And?" she continues. "Take you right down with me," he rejoins. Or as they have said throughout: "Everything five-oh, five-oh"; or in the style of *Adam's Rib*, "Equal in everything." I do not know of a more courteous bouquet of thanks from a director to his star. And yet even here Cukor has allowed himself a justifiable pride. While the camera spends in the film a disproportionate amount of time on Hepburn's physical accomplishments, a disproportionate brilliance of acting in the film is Tracy's, the good director's surrogate.

But the topic of the transfiguring of women, toward their creation or destruction, deserves to be followed thoroughly in Cukor's work. I do not know, for example, what would make it more obvious than it stands that the opening elaboration of the beauty establishment in *The Women* (1939) is an allegory of Hollywood studio film-making (which is in turn an allegory of commodity-making? or of the power of the invisible male world to turn people into commmodities? or of the power of the social as such?). And perhaps one will take Cukor's explicit treatment of Pygmalion and Galatea, in *My Fair Lady* (1964), as more or less the sheer luck of the Hollywood draw. But there is *Gaslight* (1944), in which, in a context of spiritual subjection, the theme of the good and the bad powers is incorporated into one man, and internalized by the woman, presaging madness. (*Gaslight* can be seen as Cukor's response at once to Victor Fleming's *Dr. Jekyll and Mr. Hyde* (1941) and to Hitchcock's *Shadow of a Doubt* (1943), sharing Ingrid Bergman with the former, Joseph Cotten with the latter.) More generally, and with particular gratification to me, I note from a recent late night showing of *A Double Life* (1947) that the team of George Cukor, Garson Kanin, and Ruth Gordon had placed the subject of *Othello* (whose presence of course one could hardly fail to remember if one remembered anything of the film) in conjunction with a story of the failed attempt of a pair to remarry (which I had not remembered). A more particular application of *A Double Life* to *Adam's Rib* is given in the complexity of connection it proposes between the private and the public life of a professional pair, a pair working together in the same profession. It is possible to describe the personal story of *Adam's Rib* to be the way the situation in court affects their marriage,* whereas I have claimed that the emphasis is rather on the other foot, that the marriage effects, or is expressed by, what happens in court; except, of course, that the dialectic has as yet no surcease, that the marriage requires this public expression because it is already an expression of public material. And this cycling is the unmistakable pattern between the private and the public in *A Double Life*. The pattern affirms the doubleness of human life, of human consciousness, a duplicity that collapses only with madness, or death—it is what the lunatic, the lover, and the poet imagine to compact beyond.

These remarks about other of Cukor's films are hardly more than another set of reminders about the work there is to do in putting together with my continuing definition of a genre comparable studies of the relation among genres and of the connection of both with the establishing of the *oeuvres* of the directors in play. I am led to underscore here the abutment of films of remarriage with films of the creation of the woman (or the human) by other means. Here perhaps the single greatest instance is Hitchcock's *Vertigo*, but there must also be considered the whole range of works in which the procedure of the camera can be said to inspire a creature with human life from the beginning, or to deprive a creature of it. The central case is *Frankenstein*, hardly surprising since Frankenstein was always a shadow of Pygmalion. Such a work as *The Exorcist* is significant here; in it the filmmaker virtually identifies himself with Frankenstein in synchronizing a voice for the possessed girl and in choosing eyes for her and in granting her her special powers of hurling herself and of projecting her vomit. The bad or dark side of the myth of film as furthering the creation of humanity is its revelation that our hold on our humanity is questionable, that we merely possess ourselves, inhabit ourselves as aliens. The church might accordingly involve itself not merely to oppose the Devil but to oppose a possibility it itself helped to create, in its contempt for the body. To continue this line of thought, the thought that a certain line of horror films form a shadow genre of remarriage comedies, it will be necessary to distinguish the specifically pertinent films of horror from the sets of films merely meant to terrorize us. I am speaking of horror here as I do

* As Gavin Lambert suggests in his *On Cukor* (New York: Capricorn Books, 1973), p. 200.

in *The Claim of Reason*, as a perception of the instability of the fact of human existence, its neighboring of the inhuman, the monstrous. Accordingly *The Night of the Living Dead* would fit here while, for example, *Dead of Night* would not. The comedy of remarriage is as much about separation from society, call it privacy, as the horror movie is; and as much about the establishing of civilization as the Western is. (I recall here Nevill Coghill's suggestion in distinguishing Shakespearean and Jonsonian comedy, which I cite in the reading of *His Girl Friday*, that the Shakespearean is an opposite or shadow of tragedy whereas the Jonsonian is not.)

ACCORDING TO MY UNDERSTANDING of the transfigured sexes, then, Amanda Bonner stands rebuked for invoking what Tracy Lord had called "the wrong kind of imagination,'" for harboring private motives for her public line of defense of her client, as if she has gone to such extravagant lengths to imagine the impossible because she cannot bear really to imagine the actual sordidness and dependence of Doris Attinger's life, the provocation and reduced capacity with which she did what she did. In the longest shot in the film, surely one of the longest unmoving shots in any commercial film since early silent days, as Amanda interviews Doris in jail, everything lends authority to Doris's recital—Judy Holliday's virtuoso realization of the lines, Hepburn's and Eve March's attention to them, the camera's fascination; and here Amanda is fully interested in Doris's describing herself as "watching myself, like in a dream," and indeed warns Doris not to be too sure what she meant to be doing when she shot her husband since the difference between freedom and ten years in jail is no laughing matter. Yet the first day in court Amanda challenges the first prospective juror on grounds that his beliefs are prejudicial to the idea of the equal treatment of men and women before the law, which will be her whole line of defense. The line will involve her in more or less obvious confusions, for example the idea that a woman in a trance, talking to herself, walking up and down all day long telling herself not to do anything foolish, is legally (and morally) in the same position as a man who has carefully planned to avenge himself for his wife's infidelity and who is full of the conviction that he is exercising his rights—if unwritten—as a husband. (Persons in these positions might credibly be equated on the ground that both have gone crazy.) And she rises to break up Adam's cross-examination of Doris by saying that he is "treating her as some kind of lunatic [somewhat closer to a genuine line of defense] whereas she is a fine, a healthy woman," but then in her summation she appeals to a bit of anthropology about certain descendants of the Amazons to the effect that they have treated their men tyrannically for so long that the men have become weak and incompetent through long subservience.

But my question is still how the camera has yielded its rebuke, why the experiment yields a grotesque result. And my answer, having rehearsed Cukor's preoccupation with transfiguration, is that the camera already, or naturally, captures the feminine aspect of the masculine physiognomy (and, though I am for some reason more hesitant about this, the masculine aspect of the feminine), so that the imagination required is, if properly exercised, already sufficiently prompted by the nature of projecting the human physiognomy to do what Amanda is trying to get us to will to do. Tom Ewell (that is, Mr. Attinger), when we first see him, primping a little as he hits the open air after a day of office work, swinging his body happily as he buys a paper from a newstand and walks to the subway, appears, using conventional criteria, more feminine than he does dressed in woman's clothes, which bring out (as they are realized for this film) a coarseness and masculine villainy in his features. In contrast, the two women, appearing as vulnerable young males, have discarded the insignia that made them different kinds of women and achieve before the camera a solidarity of effect that Amanda had posited for women as such when she claimed that "all women are on trial in this case." If this is the effect, Amanda's wish remains private, for the visual solidarity of the women in the face of the man's brutish villainy is not her line of defense, which requires instead that Beryl Kane be seen as a homebreaker.

(The intuition I express of the camera as revealing the reverse sexual nature of its human subjects goes with two other intuitions I have expressed about the camera. First, in connection with *The Philadelphia Story*, I spoke of the camera as revealing an otherwise invisible self, and I related this to its suggestion of the Blakean Specter and Emanation. Second, in the Foreword to the enlarged edition of *The World Viewed*, I speak of the inherent reflexiveness or self-referentiality of objects filmed and projected on a screen, of the luminosity lendable to projected objects by their "participation" [a word meant to pick up Platonic aspirations] in the photographic presenting of themselves, a presence that refers to their absence. I wish to understand as an analogue to this ontological speculation about material objects the speculation about human beings, things with consciousness, that their presence refers to their absent, or invisible, or complementary, sexuality. The reflexiveness of objects harks back, in my mind, to the earlier claim in *The World Viewed* that objects on a screen appear as held in the frame of nature, implying the world as a whole. The sexual reflexiveness of human beings would accordingly suggest the individual as expressing humanity as such, what in *The Claim of Reason* I call the internal relation of each human being with all others.)

We were led to the moment of sexual transformation as an expression of the break in the conversation between Adam and Amanda, which means their existence without their tie of marriage. As they arrive at their realization of this condition, collecting their papers and leaving the courtroom together to go their separate ways, all around them a resumption of conversation is being put into effect in a manner unthinkable for Adam and Amanda. The reporters and photographers, wanting copy to account for what they have seen, have put together not only Mrs. Attinger and her three children but Mr. Attinger with those four and then Beryl Kane with the five of them, posing the six together, all smiling, if uneasy, in a kind of family portrait. Here is a wacky, and no less genuine for being temporary, pursuit of happiness, one uncomprehended by our laws. We are not, I believe, happy with the verdict of Not Guilty, or rather we do not know what to make of it, what it should have been. But we are happier with the consequences of the verdict, seeing the children back with their parents, and we even participate a little in the contrived reconciliations. What else, until the world changes, would be a happier outcome? (A wacky, or impossible, solution to the mystery of marriage—which I found to be glanced at in the pair of closing stills of *The Philadelphia Story*—is proposed in other significant films, both inside and outside our genre. Inside, Preston Sturges, in *The Palm Beach Story* (1942) multiplies remarriages beyond necessity, or credibility; outside, in *Some Like It Hot* (Billy Wilder, 1959), Joe E. Brown accepts a male bride on the ground that nobody's perfect.)

Adam and Amanda walk out without glancing at this uproarious scene; notably, it strikes me, without a glance at the children. I take this as a comment on the fact that in risking Doris Attinger's freedom

Amanda was risking the happiness of her children, and surely it declares that an untrammeled absorption in the conversation of marriage cannot take place in the presence of children. This is the last of the major remarriage comedies. The sordidness of unequal marriage as mostly it stands, no longer checked by larger family structures, is not the joke it was fifteen years earlier in *It Happened One Night*. It is loose in the world, at society's doorstep.

I SAID THAT when the conversation between our pair resumes it will wind up on the topic of the difference between men and women. (The dialogue explicitly adds itself up this way. "Hooray for that little difference" Adam will hurriedly say, we saw, at the conclusion. And during his harangue he had said to Amanda, "We've had our little differences, and I've always tried to see your point of view." But what? This difference is different? Why? Because it is not discussible? What does that mean? Saying hooray to the little differences, in good faith, is saying hooray to exchanges about them, of them.) But how does the conversation resume? It recommences, haltingly, in the accountant's office, with some exchanges about money, which is first about what the pair have placed monetary value on in the past (intimate things, like gifts of underwear, and a silly bet on a subject that they won't share with the accountant), and then about the property they own in Connecticut, "free and clear," which introduces Adam's tears into the conversation, which then prompts Amanda to seem to take the initiative and make the proposal to leave then and there for the farm. "You mean," Tracy responds through his crocodile tears, "and see the dogs?"—leaving no room for speculation about whether children, or hence anything else, would be an acceptable substitute between these two for the conversation itself; nothing else is marriage.

Once in Connecticut, they inaugurate the trial as an acceptable topic of exchange and then at once move on to the new topic Adam says they have to discuss tomorrow, the judgeship the Republicans want him to run for. She says she's real proud of him; he replies, accepting her offer of a handshake, that he'd rather have her say that than anything. The thrill of this exchange is its putting sexuality in its place; that is, its being able to afford putting it in its place, in the confidence that it has one. In that confidence, Adam leaves the room to change into his pajamas. Amanda sees in Adam's briefcase the hat that he gave her that she gave to Mrs. Attinger that Adam ripped from Mrs. Attinger's head; starts to burn it in the fireplace, then changes her mind and puts it on; warms her toes before the fire, thinking. She calls into the next room, "Have the Democrats chosen a candidate yet?" Adam comes back slowly into the room finishing buttoning the top of his pajamas. "You wouldn't," he says, and convinces her for the moment at least by showing her how her competition there would make him make himself cry. After his demonstration, he notices that she has her hat on so he puts his on, equal to the end, ready for anything. I seem to remember learning, but I do not remember where, that the hat is an ancient symbol of liberty. I remember this in connection with the Anthony Asquith–Leslie Howard *Pygmalion* (1938), the concluding image of which is a close-up of Professor Higgins's hat, seen from behind as he has swiveled around and leaned back in his desk chair, away from the returning Eliza and from us, as if in order to display his hat, the outline of the whole circle of the brim filling the screen. (I think this image is taken over, in homage, by Cukor for his *My Fair Lady*; but here my memory is less strong than my wish.)

Since Amanda's remark upon donning the hat is to ask about the Democrats, we are entitled to take its donning as a challenge, a show of independence, while at the same time it reaccepts his gift to her. But a challenge to what? Independence from what? To and from the very fact that a conversation has resumed, and that while that is cause for happiness, that happiness is not to be presumed upon? Lines are to be drawn, or what's a conversation for? Something, I think, like that.

What more immediately precipitates her donning the hat, however—which is to say, what changes her mind about burning the hat, or burying the hatchet—is Adam's singing of the film's song off-screen, accompanying his putting on his pajamas. It is the only new element in the situation between his exit to change and her retrieving the hat from the fire. I suppose the song inspires her contention because it sounds equally like a serenade and like a song of victory.

Let us take a moment, before leaving the film, to recognize how weird a song Cole Porter has contributed to the proceedings. It begins with its title, "Farewell, Amanda," and after bidding goodbye in three further languages it requests Amanda to remember, in her new life, that wonderful night on the verandah. When a prime Cole Porter in an earlier time sang of "it" as "great fun," and as "just one of those things," he allowed the singer to hope that they would meet now and then, allowed them to survive their pleasure with some humor, some style, intact. Or when John Dryden writes "Farewell, fair Armeda" his lover understandably takes his leave to die because love is unrequited. But Kip seems to be singing his serenade to one who is herself to die ("when you're stepping on the stars above"), apparently the opportune moment for inspiring his declaration. Maybe he just means that she is moving on to higher things, and maybe he wrote the song to be sung by anyone but himself, especially by Adam. Maybe it is for him to sing on Adam's behalf, like an inverted Cyrano. He almost does. When Adam is about to leave their apartment after Amanda's homecoming to silence and his haranguing sequel, Amanda says, "Adam, don't you dare slam that door," and when he does the spiteful slamming sets off a contraption of consequences (a sort of schizophrenic Rube Goldberg machine), an image of the cunning of history, or of the logic of narrative, the last consequence of which is the turning on of a record of Kip singing his song. The record player is under pressure and you can't tell Kip's voice from Adam. So Kip half gets the wish to provide Adam's exit music, confirmation that Adam must confront him and Amanda together. And in his victory serenade Adam turns the tables, does some versifying himself. He sings the song so that it begins with the words "Hello, Amanda," and goes on to welcome her, or welcome her back, after a battle he claims was fun. In the course of his rewriting he changes "when you're stepping on the stars" to "when you're gazing at the stars," a change that brings Amanda safely down to earth, and while she must appreciate this she may also resent it a little, and resent a little the bullying talent that wrests Kip's song, which was hers, away to his own purposes. But how could she not also admire it? It is his final claim upon her, overcoming at once the brutishness of the gun and the childishness of the tears. She will yield to this achievement of gentle genitality ("Hooray for that little difference!"), but not without contesting it. After all, whose difference is it?

Hatted, as for departure, away from us, they resume their adventure of desire, their pursuit of happiness, sometimes talking, sometimes not, always in conversation.

7

THE
SAME
AND
DIFFERENT

The Awful Truth

All she will tell him, or warn him of, visiting him at his apartment, before becoming his sister, is that his ancient poem to her, which she is about to recite, will hand him a laugh.

O N certain screenings I have felt *The Awful Truth* (1937) to be the best, or the deepest, of the comedies of remarriage. This feeling may be found eccentric on any number of grounds. That I expect little initial agreement with it is registered in the qualification "on certain screenings." By the qualification I mean not only that there have been screenings on which I have not felt this way; I mean also to suggest that the experience of this film is more dependent on the quality of the individual session of screening than its companion films are. Specifically, my connection with the film, even my understanding of it, has been especially dependent, it seems to me, on the presence with me of an appreciative audience. This could mean either that my responses are less free than in other cases, requiring infectiousness and a socially inspired willingness to be pleased, to be sociable; or that my responses are more free, participating or not as they require, the film not forcing its attention upon me. Is the latter possibility really credible? It proposes an achievement of this film—that is, an achievement of its director, Leo McCarey—that transcends the comparable achievements of Frank Capra, Howard Hawks, and George Cukor. The transcendence is not, no doubt, by very much, but the surprise is that Leo McCarey should be setting the example at all for his more famous, or more prominent, colleagues. To get past what may be hardly more than prejudice here, it may help to note Jean Renoir's remark that "Leo McCarey understood people better than any other Hollywood director."* There could hardly be, from that source, higher praise.

* Reported by Andrew Sarris in *The American Cinema* (New York: E. P. Dutton and Co., 1968), p. 100.

Nor would McCarey's colleagues in the genre of remarriage themselves have been surprised at his presence. You may not find that Cukor is remembering McCarey when near the beginning of *The Philadelphia Story* Dinah says, "Nothing possibly in the least ever happens around here," three years after Aunt Patsy had said, near the beginning of *The Awful Truth*, "Nothing unusual ever happens around here." Since Dinah's line occurs in the play *The Philadelphia Story*, maybe it was just Philip Barry who was remembering Aunt Patsy a year or two later and Cukor didn't care one way or the other whether the line was retained for the film script. Or maybe the writers and directors in question were all remembering, or each work discovering for itself, a way of warning its audience, taking it by the hand as if to say, that a narrative is about to begin. ("What's happening?" asks Beckett's blind Hamm, trapped between beginning and ending.) But can it be doubted that Howard Hawks is paying homage to McCarey in all but taking over the content of the great restaurant sequence in *His Girl Friday* from the great restaurant or nightclub sequence in *The Awful Truth?* In both sequences, Cary Grant, as a group awkwardly settles itself around a table, opens a conversation with his estranged wife and with Ralph Bellamy by saying, "So you two are going to get married." Grant then quizzes them about where they will live, and elaborately pictures his wife's pleasure in getting away from the big city with its rigors of elegant shops and theaters to the peace and quiet and adventure of the West (Oklahoma City in the present film, Albany in the later). In both the woman tries to protect her new man against the onslaughts of the old and in both the conversation turns, with some relief, to a business proposition. There is also in both a moment (in *The Awful Truth* this comes not within the restaurant sequence but in the sequence that follows it in Bellamy's apartment) at which Grant, breaking up laughing as he begins reciting an intimate memory, has to be signaled off the subject by the woman. And we should note that the last night, at Grant's prospective in-laws' house, as Irene Dunne puts on her sister act, she says, in greeting the father of the family, "I never would have recognized you from his description," thus preparing the way for Walter's initial words to Bruce, "Hildy, you led me to expect a much older man."

If *The Awful Truth* does have a certain specialness, perhaps this is to be attributed less to its director than to some special place it occupies in the genre of remarriage. It is the only member of the genre in which the topic of divorce and the location in Connecticut are undisplaced, that is, in what one is most likely to take as their natural places; in which the pair's story both opens with the former and closes at the latter. But how do we know that this kind of natural or straight account is so important, more important, say, than the fact that in this film the woman's father is not present but is replaced by someone called the woman's "Aunt"? Besides, if genre itself were decisive, Hitchcock's *Mr. and Mrs. Smith* (1941), which works brilliant variations within the genre, would have more life for us than is to be derived from its somewhat cold comforts. Any answer having to do with the depth of participation in the genre must invoke a director's authority with the genre, his nativeness or subjection to it, the director and the genre knowing how to get the best from one another.

And this must mean, according to our understanding of this genre, knowing how to take a woman most deeply into the forces that constitute the genre, which in turn means finding a woman, and finding those qualities in a woman, in whom and in which those forces can most fully be given play. Here is a place we come unprotectedly upon the limitation of criticism by the fact of something that is called personal taste. About *It Happened One Night* I said that its appreciation depended on a certain acceptance of Claudette Colbert; but my sense of *The Awful Truth* is that if one is not willing to yield to Irene Dunne's temperament, her talents, her reactions, following their detail almost to the loss of one's own identity, one will not know, and will not care, what the film is about. Pauline Kael, for instance, in her Profile of Cary Grant has this to say about Irene Dunne in *The Awful Truth*: "though she is often funny, she overdoes the coy gurgles, and that bright toothy smile of hers—she shows both rows of teeth, prettily held together—can make one want to slug her."* Whatever the causes of this curious response, it disqualifies whatever she has to say as a response to *The Awful Truth*.

IT IS, I believe, particularly hard to recall the sequence of events that constitute the film; and since I am going to take something like this difficulty to be internal to McCarey's achievement in it, it will help to

* *The New Yorker*, July 14, 1975. Reprinted in *When the Lights Go Down* (New York: Holt, Rinehart and Winston, 1980), p. 7.

summarize its main segments. (1) In a prologue, at the Gotham Athletic Club, Jerry Warriner (Cary Grant) is about to get a sun-lamp treatment sufficient to make it appear that he's spent the last two weeks in Florida, "even if it takes all afternoon." He is speaking to a passing acquaintance with a squash racket: "What wives don't know won't hurt them." And he adds, "And what you don't know won't hurt you." He invites the acquaintance to come home with him later on for protection, I mean for drinks. (2) Entering the house with this, and other acquaintances, Jerry discovers that his wife is not at home. He invents the explanation that she's at her Aunt Patsy's place in Connecticut, an explanation which collapses when Aunt Patsy walks in looking for her. Lucy Warriner (Irene Dunne) enters, in evening dress, followed by Armand Duvall, her singing teacher, it emerges, with a story about chaperoning a dance and then on the road back having the car break down miles from nowhere, and spending the night at a very inconvenient inn. Jerry mockingly pretends to believe the story and is complimented by Armand for having "a continental mind." The guests take the cue to leave, Jerry says his faith is destroyed, Lucy says she knows what he means and tosses him a California orange that he had brought her as from Florida. She says he's returned to catch her in a truth, to which he responds by calling her a philosopher. He gives a speech which includes the lines: "Marriage is based on faith. When that's gone everything's gone." She asks if he really means that and upon his affirming it she telephones for a divorce. (3) Her lawyer, on the phone, repeatedly tells Lucy not to be hasty, that marriage is a beautiful thing; he is repeatedly interrupted by his wife asking him why they have to be interrupted, whom he repeatedly invites, each time covering the phone, to shut her mouth. (4) In divorce court, Mr. Smith (the dog Asta) is tricked by Lucy into choosing to live with her. Jerry asks for visiting rights. (5) Aunt Patsy wants to get out of the apartment she and Lucy have taken and have some fun tonight for a change. Lucy objects that they haven't an escort. Aunt Patsy stalks out and comes back with their neighbor from across the hall, Dan Leeson (Ralph Bellamy). Jerry appears for his visiting time with Mr. Smith, whom he accompanies at the piano. The others leave. (6a) Dan's mother warns him in general about women and in particular about that kind of woman; (6b) Aunt Patsy warns Lucy against acting on the rebound, pointing out to her that her toast is burning; (6c) 6a continued; (6d) 6b continued. (7) In a nightclub Jerry's friend Dixie Belle sings and enacts "My Dreams are Gone with the Wind." On each recurrence of the title line air jets from the floor blow Dixie Belle's flowing skirt up higher and higher—she finally gives up trying to hold it down. On meeting Dixie Belle, Lucy had said, with some surprise (presumably given her view of Jerry's taste in women), that she seems like a nice girl. When Jerry corrects Dan's impression that Lucy dislikes dancing, Dan, from whom we learn that he is a champion dancer, takes her onto the floor. The music changes and Dan is moved to take over the floor with his champion jitterbugging. Jerry so thoroughly enjoys Lucy's taste of country life that he tips the orchestra to repeat the same number. Jerry pulls up a chair to the edge of the dance floor, sits legs crossed, his arms draped before him carelessly, perfectly, fronting the dancers and the camera, looking directly at the world with as handsome a smile as Cary Grant has it in him to give, in as full an emblem of the viewer-viewed, the film turned explicitly to its audience, to ask who is scrutinizing whom, as I know in film. I think of it as a hieratic image of the human, the human transfigured on film. This man, in words of Emerson's, carries the holiday in his eye; he is fit to stand the gaze of millions. Call this the end of Act One. (8a) Lucy and Dan at the piano in his apartment make a duet of "Home on the Range"; (8b) Jerry enters to discuss their business deal about a mine; (8c) Dan's mother comes in with gossip about Lucy; Jerry sort of clears her name with a speech of mock gallantry, exiting on the line, "Our marriage was one of those tragedies you read about in the newspapers," but Maw is still not satisfied, whereupon Lucy retreats to let her and Dan sort the matter out alone. (9) Lucy returns to her apartment to find Jerry there, rewarding himself with a drink for having, he says, given her that swell reference; she haughtily refuses an offer of financial help from him, and laughs heartily as the piano top falls on his hand. As they walk toward the door for Jerry to leave, Dan knocks. She opens the door, concealing Jerry behind it. Dan apologizes for his mother's suspicions and insists on reading Lucy a poem of love he has written for her. As he embarks on it Jerry from behind the door prods Lucy to laughter with surreptitious pencil jabs in her ribs. The phone rings, just the other side of Jerry. Lucy answers; we are shown by an insert that it is Armand; Lucy asks whoever it is to wait and puts down the phone; as she crosses back past Jerry to complete her exchange with Dan, behind her back Jerry picks up the phone and learns who is on the other end.

Lucy gets rid of Dan by giving him a kiss; he departs noisily. Lucy makes an appointment into the waiting phone, handed to her by Jerry, for three o'clock the next afternoon, explaining to Jerry after she hangs up that it was her masseuse. Jerry finally leaves, saying he's just seen a three-ring circus. The situation prepares for the juggling of farce. (10) At three o'clock, evidently the next afternoon, Jerry forces his way into Armand's apartment only to discover Lucy singing for a musicale. (11) The farce erupts as Mr. Smith fetches Armand's hat for Jerry, whom it doesn't fit, try as he will. The two men find themselves in the same bedroom, Armand to avoid Jerry, Jerry to avoid Dan and Maw who have come together this time to apologize again. From the bedroom the two men dash across the living room past the assembled others and out the door. Lucy had written a letter to Dan telling him that she was still in love with Jerry and had asked Aunt Patsy to deliver it. Dan says, a moroser if wiser man, "I've learned a lot about women from you, Lucy; I've learned that a man's best friend is his mother." As he and his best friend start their exit, Aunt Patsy takes Lucy's letter from the mantle and delivers it: "Here's your diploma." Call this the end of Act Two. (12a) Mr. Smith barks at the society page of the newspaper Lucy is reading; it says that Jerry and Barbara Vance are to be married as soon as his divorce is final, which incidentally is today; (12b) The newspaper comes alive in a montage of Jerry and Barbara's whirlwind romance, which mostly consists of their attending or participating in society sports events; a sequence that reads like the society segment of, say, a *Movietone News*. (13) In Jerry's apartment, to say goodbye on the eve of their final decree, Lucy recites a poem written in another time for her by Jerry. She introduces it by saying, "This will hand you a laugh," but neither of them is tickled in the ribs. They sample some champagne that the life has gone out of; evidently they are unable to celebrate either divorce or marriage. To account for Lucy's presence when she answers a phone call from Barbara Vance, Jerry invents the tale that his sister is visiting from Europe, then after a pause explains that she can't come over with him tonight because she's busy and anyway is returning to Europe almost immediately. Lucy says he's slipping. (14) That night, at the Vance house, Lucy interrupts a flagging family occasion with a vulgar display as his low-down sister. She claims to be a nightclub performer and shows them how with Dixie Belle's "Gone with the Wind" ("There's a wind effect right here but you will just have to use your imagination"). Jerry joins her on her exit from the song and dance, and (15) they drive to their conclusion in Connecticut.

THAT THERE IS NOT a dull scene in the film is less important a fact, or less surprising given its company, than that there is no knock-out scene, nothing you might call a winning scene, until perhaps the end of the two-men-in-a-bedroom farce, which in my outline I figured as the end of Act Two; but possibly the preceding recital sequence can be taken so, or perhaps the sequence preceding the recital, as Jerry pokes Lucy in the ribs to laugh at Dan's poem. Even if you consider what my outline figured as the end of Act One as a winning scene, this night club sequence takes place much later in its film than, say, the restaurant sequence of *His Girl Friday* in its film, which already followed several instances of knock-out business. We have in this absence of a certain kind of scene the beginning of an explanation of the particular achievement of this film—*if*, that is to say, one regards this film as a serious achievement. Speaking of an absence in this regard is putting negatively a virtue that, put positively, empowers the presenting of an unbroken line of comic development, a continuous unfolding of thought and of emotion, over a longer span than is imagined in the companions among the genre of comedy in which we are placing the film. I understand the point of the achievement to be the tracking of the comedic to its roots in the everyday, to show the festival to which its events aspire to be a crossroads to which and from which a normal life, an unended diurnal cycle, may sensibly proceed. I want to spell out this perception a little further now, if more or less abstractly, as a kind of gauge of this film's role in the genre of remarriage.

The diurnal succession of light and dark takes the place in these films of the annual succession of the seasons in locating the experience of classical comedy. The point is to show that the diurnal, the alternation of day and night, and in the city, mostly sheltered from the natural seasons (as in a film studio), is itself nevertheless interesting enough to inspire life, interesting enough to be lived happily; lived without, one may say, outbreaks of the comic, as if there is no longer a credible place from which our world can be broken into; that is, no communal place, no place we have agreed upon ahead of time. An answer is being given to an ancient question concerning whether the comic resides fundamentally in events or in an attitude toward events. In claiming these films to enlist on the side of attitude here, I am assuming that sanity requires the recognition of our dependence upon events, or happenstance. The suggestion is that happiness requires us not to suppose that we know ahead of time how far, or where, our dependence on happenstance begins and ends. I have had occasion in speaking of the career of Othello to invoke Montaigne's horrified fascination by the human being's horror of itself, as if to say: life is hard, but then let us not burden it further by choosing tragically to call it tragic where we are free to choose otherwise. I understand Montaigne's alternative to horror to be the achievement of what he calls at the end a gay and sociable wisdom. I take this gaiety as the attitude on which what I am calling diurnal comedy depends, an attitude toward human life that I learn mostly from Thoreau, and partly from Kierkegaard, to call taking an interest in it. Tragedy is the necessity of having your own experience and learning from it; comedy is the possibility of having it in good time.

(Should someone take the ideas of attitude and of perspective here as being matters of some known element of psychology, say of some particular feeling or matter of will, it may help to say that attitude and perspective enter as well into the constitution of knowledge, the constitution of the world. The difference between taking a statement as true a posteriori or as true a priori can be said to be a difference in the attitude you take toward it. When the hero of *Breathless* says "There is no unhappy love," he is not, as some may be, leaving the matter open to question, to evidence; for him it is knowledge a priori; you may say a definition. One wants to say here: it is a truth not necessary in all imaginable worlds but necessary in *this*—I mean in *my*—world: "When I love thee not, chaos is come again"—at that moment there will be no world, things will have gone back to before there is a world. And attitude and perspective, and I suppose something like the same division of attitude and perspective, are at play in the distinction between the factual and the fictional. The question is again how a matter gets *opened* to experience, and how it is *determined* by language or, let us say, by narration. The truths of arithmetic cannot be more certain than that Hamlet had a doublet and wore it all unbraced. Ophelia's word for it *cannot* be doubted. Some who concern themselves with the problem of fiction may be making too little of the problem of fact.)

THE AWFUL TRUTH

"The tracking of the comedic to its roots in the everyday." This is my formulation of the further interpretation of the genre of remarriage worked out in *The Awful Truth*. I intend it to account for several features of the genre that differentiate it from other comic forms.

For example, the stability of the conclusion is not suggested by the formula "they lived happily ever after" but rather requires words to the effect that *this* is the way they lived, where "this" covers of course whatever one is prepared to call the conclusion of the work but covers it as itself a summary or epitome of the work as a whole. (In Chapter 3 I express this density of the conclusion by speaking of its aphoristic quality.) There is no other life for them, and this one suffices. It is a happy thought; it is this comedy's thought of happiness.

Again, I have pointed several times to the absence, or the compromise, of the festival with which classical comedy may be expected to conclude, say a wedding; I have accounted for this compromise or subversion by saying variously that this comedy expects the pair to find happiness alone, unsponsored, in one another, out of their capacities for improvising a world, beyond ceremony. Now I add that this is not to be understood exactly or merely as something true of modern society but as something true about the conversation of marriage that modern society comes to lay bare. The courage, the powers, required for happiness are not something a festival can reward, or perhaps so much as recognize, any longer. Or rather, whatever festival and ceremony can do has already been done. And wasn't this always true? In attacking the magical or mechanical view of the sacraments, Luther says, "All our life should be baptism." I once took this as a motto for romantic poetry.* I might take a variation of it as a motto for the romance of marriage: all our life should be festival. When Lucy acknowledges to Aunt Patsy her love for Jerry after all, what she says is, "We had some grand laughs." Not one laugh at life—that would be a laugh of cynicism. But a run of laughs, within life; finding occasions in the way we are together. He is the one with whom that is possible for me, crazy as he is; that is the awful truth.

"Some grand laughs" is this comedy's lingo for marriage as festive existence. The question, accordingly, is what this comedy means by laughter. Whatever it means it will not be something caused and pre-

---

* "A Matter of Meaning It," in *Must We Mean What We Say?*, p. 229.

vented by what we mostly call errors. This is a further feature in which the comedy of remarriage differs from other comedy.* The obstacles it poses to happiness are not complications unknown to the characters that a conclusion can sort out. They have something to learn but it cannot come as news from others. (Nor is our position as audience better in this regard than that of the characters. To the extent that the effect of classical comedy depends on a sense of our superiority to comic characters, the comedy of remarriage undermines that effect. We are no more superior to these characters than we are to the heroes and heroines of any adventure.) It is not a matter of the reception of new experience but a matter of a new reception of your own experience, an acceptance of its authority as your own. Kierkegaard wrote a book about our having lost the authority, hence so much as the possibility, of claiming to have received a revelation.** If this means, as Kierkegaard sometimes seems to take it to mean, the end of Christianity, then if what is to succeed Christianity is a redemptive politics or a redemptive psychology, these will require a new burden of faith in the authority of one's everyday experience, one's experience of the everyday, of earth not of heaven (if you get the distinction). I understand this to be the burden undertaken in the writing of Emerson and of Thoreau; doubtless this is a reason it is hard to place them in a given field. One might take the new burden of one's experience to amount to the claim to be one's own apostle, to forerun oneself, to be capable of deliverances of oneself. This would amount to an overcoming of what, in *The Claim of Reason*, I call the fear of inexpressiveness.† Here is a form in which art is asked to do the work of religion. Naturally this situation makes for new possibilities of fraudulence, among both those who give themselves out as apostles and those who think of themselves as skeptics.

It is centrally as a title for these three features of diurnal comedy, the comedy of dailiness—its conclusion not in a future, a beyond, an ever after, but in a present continuity of before and after; its transformation of a festival into a festivity; its correction not of error but of experience, or of a perspective on experience—that I retain the concept of remar-

---

* I call attention here to Harry Levin's rich Introduction to the Signet edition of *The Comedy of Errors.*

** An essay of mine about that book, "Kierkegaard's *On Authority and Revelation,*" appears in *Must We Mean What We Say?*

† See, for example, pp. 351, 473.

riage as the title for the genre of films in question. The title registers, to my mind, the two most impressive affirmations known to me of the task of human experience, the acceptance of human relatedness, as the acceptance of repetition. Kierkegaard's study called *Repetition*, which is a study of the possibility of marriage; and Nietzsche's Eternal Return, the call for which he puts by saying it is high time, a heightening or ascension of time; this is literally *Hochzeit*, German for marriage, with time itself as the ring. As redemption by suffering does not depend on something that has already happened, so redemption by happiness does not depend on something that has yet to happen; both depend on a faith in something that is always happening, day by day.

Thus does the fantasy of marriage being traced out in these chapters project a metaphysic, or a vision of the world that succeeds the credibility of metaphysics. It was only a matter of time, because as the fantasy becomes fuller and clearer to itself it poses for itself the following kinds of question. What must marriage be for the value of marriage to retain its eminence, its authority, among human relations? And what must the world be like for such marriage to be possible? Since these are questions about the concept as well as about the fact of marriage, they are questions about marriage as it is and as it may be, and they are meant as questions about weddedness as a mode of human intimacy generally, intimacy in its aspect of *devotedness*.

This recent conjunction of ideas of the diurnal, of weddedness as a mode of intimacy, and of the projection of a metaphysics of repetition, sets me musing on an old suggestion I took away from reading in Gertrude Stein's *The Making of Americans*. She speaks, I seem to recall, to the effect that the knowledge of others depends upon an appreciation of their repeatings (which is what we are, which is what we have to offer). This knowing of others as knowing what they are always saying and believing and doing would, naturally, be Stein's description of, or direction for, how her reader is to know her own most famous manner of writing, the hallmark of which is its repeatings. The application of this thought here is the suggestion that marriage is an emblem of the knowledge of others not solely because of its implication of reciprocity but because it implies a devotion in repetition, to dailiness. "The little life of the everyday" is the wife's description of marriage in *The Children of Paradise*, as she wonders how marriage can be a match for the romantic glamour of distance and drama. A relationship "grown sick with obligations" is the way Amanda Bonner describes a marriage that cannot maintain reciprocity—what she calls mutuality in everything. (This is a promissory remark to myself to go back to Stein's work. But the gratitude I feel to it now should be expressed now, before looking it up, because it comes from a memory of the work as providing one of those nightsounds or daydrifts of mood whose orientation has more than once prevented a good intuition from getting lost. This is not unlike a characteristic indebtedness one acquires to films. It is just such a precious help that is easiest to take from a writer without saying thanks— and not, perhaps, because one grudges the thanks but because one awaits an occasion for giving it which never quite seems to name itself.)

As the technical, or artistic, problem of the conclusion of the members of the genre of remarriage is that of providing them with epitomizing density, the artistic problem of the beginning of *The Awful Truth* is to preserve its diurnal surface, to present comic events whose dailiness is not interrupted by comic outbreaks but whose drift is toward a massive breakthrough to the comic itself as the redemption of dailiness, a day's creation beyond itself. The risk of such a structure is dullness; it must open an accepting tameness, domesticity, as one pole of the comic (as Mr. Pettibone does). The reward of the structure is scope, the distance it gets in discovering its conclusion. One might picture the narrative structure as preparations followed by surprises (like a chess game); or perhaps as sowing followed by reaping. This would leave out the fact that the sowing is a sequence of reapings (or of surprises too mild individually to be noticed as such), increases of interest, of a willingness to be pleased, say to be civilized; and the reapings a sequence of discoveries whose originality cannot be thought of as sown, unless perhaps as dragon's teeth.

WHAT I CALLED the abstractness of these claims has its own interest, but it is useful here to the extent that the claims alert us to points of the film's concreteness that we might otherwise slip past. This should become assessible as we now go on to follow certain lines of force through the film.

The Awful Truth is the only principal member of the genre of remarriage in which we see the central pair literally take their own marriage to court (sequence 4). The point of the sequence is to dramatize the dog's role as the child of the marriage (though really he is its muse, since a squabble over who was to buy him was the thing that, according to Lucy's sworn testimony, precipitated their marriage). How funny is it when the dog is asked to choose which of the pair he wants to live with, and when Lucy tricks him with Jerry's home-coming present of a rubber mouse into "choosing" her? About as funny as the idea that these people do not know what it means that they happened to get married as if to make a home for a dog, and that a court of law is no more capable of telling them whether their marriage has taken, or is worth the taking, or else the leaving, than it is of determining reasonably the custody of a dog.

You learn to look, in a McCarey scene, for the disturbing current under an agreeable surface. He has the power to walk a scene right to that verge at which the comic is no longer comic, without either losing the humor or letting the humor deny the humanity of its victims. (Not for nothing is he the director of the Marx brothers as well as of *Love Affair*, 1939, also with Irene Dunne.) Chaplin and Keaton cross the verge into pathos or anxiety, as if dissecting the animal who laughs, demonstrating the condition of laughter. Hawks crosses the verge without letting you stop laughing; as he does in his adventures. What do people imagine when they call certain film comedies "madcap"? Do they imagine that a virgin's burning brain is in itself wildly comic, or particularly so if she is free enough, that is, if her father is rich enough, that no magic can stop her from laughing, that is, from thinking and trying not to think? Aunt Patsy will call Barbara Vance a "madcap heiress." This seems a tip from McCarey that calling his comedy "madcap" would be about as useful as taking the humorless, conventional, all but nonexistent Barbara Vance to be the heroine of his film. (A tip reinforced the next year by Hawks, or one of his sources, in naming the heroine of *Bringing Up Baby* Susan Vance.) This is, in any case, not exactly our problem since the women we are following are on the whole to be understood as married. Now for the last time: What is comic about that?

(Before proceeding, I note a further point about the occurrence of the epithet "madcap heiress" in this film. It is what sets off the thing I called in my outline of the film's sequences, at sequence 12b, the transformation of the newspaper photo into an installment of *Movietone News*. The implication is that the invention of stories about madcap heiresses is the work of scandalmongers, of gossip columnists or of movie reviewers, not the business of serious comic films. Accordingly when Jerry describes, for the benefit of Dan's mother, his and Lucy's marriage as "one of those tragedies you read about in the newspapers," he may be taken to mean that something newspapers call a tragedy is as likely as not to be what newspapers would make of it, unless perhaps they made of it a madcap comedy. These are further moments in the vicissitudes of the image of the newspaper that constitutes a feature of the genre of remarriage. While I have noted major occurrences of this feature, or its equivalent, in all but one of the films of our genre [I have not looked for an explanation of its absence in *Bringing Up Baby*], I have not given the attention it deserves to formulating the significance of their interrelations.)

In the opening sequence, what does Jerry mean by "what wives don't know" and by "and what you don't know"? It is definite that he has been away from home for two weeks and that he has told his wife he was to be in Florida. But nothing else is definite. For all we know what doesn't hurt his acquaintance because he doesn't know it is that there is nothing to know, of the kind the acquaintance suspects. Maybe for some reason Jerry is less interested in the fact of philandering than in the possibility of it, that what is important to him is not the cultivating of dalliances but the cultivating of a reputation for them. It seems only mildly awkward for him when his wife shows him, in the following sequence, that she knows he hasn't been in Florida. Anyway why was Florida safer than New York? His being caught in a lie is less relevant to their ensuing agreement to divorce than her being, as she puts it, caught in a truth. Why? What is so awful about the truth that nothing happened? And why would a married man find it more important to seem unfaithful than to be so? Perhaps it is his way of dramatizing his repeated philosophy that "Marriage is based on faith. If you've lost that you've lost everything"; his way of testing her faith, a test he himself seems miserably to fail. Is he projecting his guilt upon her? Is he withholding his innocence from her? What we know is that he is hiding something and that he is blaming her for something.

I have noted that divorce is asked for by asking to be free. If what Jerry is trying to establish is what we might call the freedom of marriage, then his complex wish for reputation is logical. All that freedom requires is, so to speak, its own possibility. As long as he *can* choose he *is* free—free for example to choose faithfulness. This would be creating a logical space within marriage in which to choose to be married, a way in which not to feel trapped in it. But it turns out that this space will have to be explored by Jerry in the way our genre dictates, by his choosing to remarry, to begin again.

An effort at freedom is mocked in the ensuing scene (sequence 2) when continental Armand repeatedly praises Jerry for having a continental mind. That he hasn't any such thing where his wife is concerned is evident enough; but it is a way of describing the reputation we have surmised him to want to establish for himself. Lucy picks up on this theme when she says to Armand, who offers to stay to protect her from Jerry's accusations, "It's all right. American women are not accustomed to gallantry." The film is announcing itself to be both in and out of a tradition that includes French farce and Restoration comedy, which means declaring its territory to be America and to be cinema. Freedom in marriage is not to be discovered in the possibility of adultery, which thus becomes unusable for comedy; it becomes either irrelevant or else the stuff of melodrama. But Jerry could not be imagined to be, however obscurely, declaring his freedom in marriage apart from imagining him to be responding to some sexual contention between himself and Lucy. That sexuality is under contention means both that sexual satisfaction is a reasonable aspiration between them, and that divorce is a reasonable, civilized alternative. (It is, for example, made explicitly clear that divorce is economically feasible for the woman.) But what this contention covers, what this dragon's tooth will produce, only time can tell.

McCarey is in a position to declare a distance from French farce because he shows himself, in what Jerry calls "that two-men-in-a-bedroom farce," expert, where required, in putting farce on the stage—or rather, in staging it for the screen. That McCarey's farce is made not for the stage but for film is in effect stated by giving the dog a pivotal, independent role in its choreography, with trickier and more irreversible business (for example, a mirror worked by the dog away from a wall to crash at a farcically timed moment) than you could count on for each performance of a play. Mr. Smith, the muse of the marriage, seems here to be preventing its putting itself back together. But he is really saying only what the film is saying, that the marriage will not go back together until it goes further, until, that is to say, the pair's conversation stops putting an innocent bystander into the woman's bedroom. There are always bystanders, one as innocent as the next. Here farce is the name of that condition of a life whose day and night must be kept from touching, which apprehends the approach of truth as awful. Not being tragic, irreversible, it is here a condition of which the right laugh would be the right cure.

In comparison with the brilliance of the farce sequence (sequence 11), the little sequence of Lucy and Dan singing "Home on the Range" (7a), with which Act Two opens (according to my figures), can seem so tame, or thin, as to give no support to thought at all, or for that current of disturbance that I have said McCarey keeps in circulation. But I find the little sequence equal to the farce—not equal in the virtuosity of its business, which is next to nothing, but in its compression of concepts.

The woman accompanies herself and her suitor as together they sing a colossally familiar tune, one no American could fail to know, not something folk but something folksy, a favorite butt of sophisticated society. Dan Lesson has virtually stepped out of its shell. It is on this note that these Americans can meet, that any American can meet any other; they cannot therefore merely despise it. The woman does not despise it, the man might just mean it, as it stands; this would be for a woman of her gallantry sufficient reason not to despise it. She even perceives the genuine longing, a moment of originality, in the song's variation of its opening five note pattern at the words "Home, home on the range"; the itensity in this variation, both in the words and in the music, is the occasion for her departure into harmonizing—a departure Dan cannot bear up under. (To check the rightness of her departure, say the words without the tune. To gauge the song's originality, or passion, compare it with "The Man on the Flying Trapeze," the song of *It Happened One Night*, which opens with the identical configuration of five notes, which then repeat on each recurrence without variation.) When she compliments them on their performance he replies, "Never had a lesson in muh life. Have you?" These sentences pretty well seal the man's fate, whatever she and Jerry will be able to work out. This man does not know who this woman is, he does not appreciate her; these things follow from his not appreciating her voice and her attitude toward her voice. That exchange about lessons is a gag based on the knowledge that Irene Dunne is a singer, a piece of knowledge no one who knew anything about her could have failed to know. The initial point of the gag is its satisfaction of the demand of the genre that each member of it declare the identity of the flesh and blood actress who plays its central female character. The consequent point of the gag is to establish that in the fiction of this film as well the identity of the character played by this woman, the one called Lucy Warriner, is also of someone identified with and by her voice. This becomes increasingly pertinent.

In the scene preceding the duet, in the nightclub, Dan's hesitation in recognizing Dixie Belle's self-evidently southern accent as a southern accent need not be taken to show his unrelieved stupidity. He is intelligent enough to have made and preserved a lot of money and intelligent enough to have fallen in love with this particular woman and to have asked her, as they and Aunt Patsy leave Jerry and Mr. Smith alone in her apartment that first night, whether she still cares for that fellow, applying a parable from his experience with perfect accuracy: "Down on my ranch I've got a red rooster and a little brown hen. They fight a lot too. But every once in a while they stop fighting and then they can get right friendly." His hesitation over Dixie Belle's accent is directly a proof solely that he has no ear. His reaction after his duet with Lucy just goes to show that in the world of these films this lack of ear is fatal.

And what is that terrible American pride in never having had a lesson? Is it different from taking pride in any other handicap? I suppose it is no worse than taking pride in having had a lot of lessons, or in being free of handicaps. Dan has an American mind. His ideology of naturalness with respect to human or artistic gifts is to be assessed against a continental ideology of cultivation (call it pride in lessons), attitudes made for one another; and assessed along with his ideology of exploitation of the gifts of nature (expressed in the next scene by his declared experience and hence training in making the mineral contents of his land holdings pay off). His pride in his empty mediocrity with art serves to underscore his deafness to the fact that the woman is accomplished and, moreover, that her attitude toward her accomplishments has a particular humor about it, not making too much of her natural extension of a capacity most people have, secure in the knowledge that it can—for many a normal person, ones without handicaps of the ear—provide pleasures. Her attitude, the pleasure she takes in her gifts, is as internal to the pleasures they give as Fred Astaire's acceptance of his own virtuosity is to the pleasure it gives, not making too little of his supernatural extension of a capacity most people have. Dunne and Astaire share this signal mark—making neither too much nor too little of something—of sophistication.

"Home on the Range" is, finally, and not altogether surprisingly, about home, or rather about a yearning to have a home. One might have doubted whether this is pertinent to its presence in a genre which is so centrally about the finding, or refinding, of a home; but I assume this doubt is allayed in recognizing that the other featured song in *The Awful Truth*, "My Dreams are Gone With the Wind," is also about this yearning, or dream: "I used to dream about a cottage small, a cottage small by a waterfall. But now I have no home at all; my dreams are gone with the wind." (Not surprisingly the man's idea of home invokes open spaces, the woman's invokes closed. But it really is a home on the range Dan Leeson has made for himself, not sheerly taken the range itself as home, however much the American male's inheritance of Huck Finn may fantasize this possibility, this way of taking the song.) As said, the singing of the male song, or rather the man's responses to the singing of it, places irrecoverable distance between the two who sing it; the singing of the female song has the opposite effect.

BEFORE CONSIDERING that effect let us loop back and collect the instances of singing throughout the films of remarriage we have been reading. It seems a firm commitment of the genre to make room for singing, for something to sing about and a world to sing in.

His Girl Friday is the only exception to this rule; it is part of its blackness to lack music almost entirely. I recall only a few bars of Hollywood up-sweep during its last seconds, startlingly breaking the musical silence, as if to help measure the abnormality of this depicted world one last time before helping to clear the theater. Most recently, concerning *Adam's Rib*, we spoke of a man using a song as his capping claim upon a woman. In *Bringing Up Baby* the pair sing "I Can't Give You Anything but Love, Baby" to soothe the savage breast of a leopard (whatever that is). In *The Lady Eve* it is the man's father who sings and whose song helps lend him the authority he will require to affect the conclusion. It happens, however, that his son, our hero, is whistling his father's tune about filling the bowl until it doth run over as he awaits his heroine on the deck the morning after the night of their first encounter. Since the father's song occurs in the film much later than the son's whistling it is virtually impossible to note the coincidence on a single viewing.

"The Man on the Flying Trapeze" is variously a good song for *It Happened One Night*. Its folk song alternation of verse and refrain allows Capra to get from it not only a general occasion for an expression of social solidarity, but a specification of this solidarity as one in which individual (taking the verse) and society (giving the refrain) exchange celebratory words with one another in harmony and with pleasure. It is also to the point that the song is about the spoiling of innocence and domesticity by male glamour and villainy. It is further to the point that Shapely's seedy, unsocial villainy is expressed as his leaving himself out of the song. It is while the society of the bus is cheerfully affirming its solidarity that Shapely discovers Ellie's photograph in a newspaper and looks back knowingly over his seat at her and Peter singing. (That the value of singing, for Capra, lies in its moral or social power and not in its isolated aesthetic power, that is to say, in what Capra understands as isolated, is emphasized by the figure of the road thief, who will sing more or less incessantly and whose singing is not without a certain aesthetic standing. But his is a narcissistic kind of vocalizing, not a way of casting his lot with others; it is a form of emotional theft.) Then why is it just when the bus driver is himself drawn into the song and lifts his hands from the wheel momentarily to begin the "Oh-h-h" of the refrain, that the bus of state skids off the road (into a depression)? What happens next is that the mother is discovered to have fainted from hunger. Is the idea that society has skidded *because* it, or its leadership, was blindly drawn away from attending to business? (The great binge of the twenties followed by the morning after of the thirties—a view of the Depression presented, perhaps itself mocked, in a New Yorker cartoon of the period which shows a society party in full swing on an airplane which is about to crash, into a mountain.) Or is it that the solidarity is compromised by those who are left outside the song of society—ones too poor to sing, whom private good will must pause to succor, and ones too cowardly and self-centered to join in song, spectators of society, not participants in it, who will have to be scared off? (Peter gives the child of the mother who has fainted his last ten bucks, or rather Ellie gives it, assuming that there is more where that came from; still she gives it. And Peter scares off Shapely with another yarn.)

One might speak of this singing as over-determined. A reason not to speak so may seem to be that Freud's concept of over-determination describes the formation of mental phenomena, for example of symptoms or of images in a dream, where the point is that just *this* symptom or image has occurred and not something else. Whereas what? The song in this film might have been different and differently sung and differently placed? But the fact that just *it* is here, where and as it is here, is what I wish to account for. Over-determination seems a good name for the formation of its appearance since the concept does not prejudge how much in the appearance is the result of intention and how much of the genre, how much is the result of specific function and how much of general structure.

Still, among the determinants of singing throughout our genre, I emphasize singing's special relation to the man, as though the man's willingness to sing, or readiness to subject himself to song, is a criterion of his fitness for the woman. And I emphasize the characteristic sound shared by "The Man on the Flying Trapeze" and "Home on the Range" with other songs in three-quarter time that invoke the social pleasures of the out-of-doors or of popular entertainments, songs such as "Bicycle Built for Two" and "Take Me Out to the Ball Game." In coming from an era essentially earlier than the time of these films, in constituting perhaps the first sound of the universally and persistently popular American song, these songs establish what Americans are apt to think of as the popular in song: the ground, I was saying, on which any American can meet any other. This force is most surprisingly confirmed when in *The Philadelphia Story* Dexter is finally moved to sing. That he must sing, enter his claim that way, is explicitly and locally established by his having to claim Tracy at the hands of Mike, who has already established his claim by singing. Dexter is moved to song by the ecstasy of seeing George depart. He rushes to a decorated wedding table and lifts its candles out of their arrangement one by one to shake them as if ringing tuned bells. Thus accompanying himself, invoking peals of bells, he sings the opening phrase and a half of "The Loveliest Night of the Year," another three-quarter time tune in the major mode associated with the circus. That it is this sound of the popular and the

association with a popular form of entertainment that is what is pertinent is registered in Dexter's singing not with words but with universal
dah-dah-dah's. Whether his ecstasy is that of a child going to the circus
or of a man getting rid of a clown, it is unimportant to decide; and
surely it heralds some tightrope walking. The imaginary ringing of bells
seems to be what then sets off the wedding music to begin the closing
festival. That Dexter's tune is popular where Mike's ("Over the Rainbow") had been sophisticated plays into the hands of the American or
national aspirations I was pressing toward the end of the discussion of
The Philadelphia Story, both by having the heroes from different classes
equal in song, even possibly reversed in their allegiances to sophistication, and by suggesting that film is the mode of modern entertainment
in which the distinction between the popular and the learned or the serious breaks down, incorporating both.

ONLY IN *The Awful Truth* among the members of our genre does the
woman sing for the man, for his pleasure and for his commitment.
(Marlene Dietrich and Mae West have sung for these reasons, but then
the singing was not for former husbands, and probably not for prospective ones either. This is part of the point of Irene Dunne's song to her
man, as will emerge. In "More of *The World Viewed*" I speak of Dietrich
and West, along with Garbo, as "courtesan figures" who seem "to
triangulate the classical possibilities outside of marriage"; p. 206.)
Hence only here can the man show his inhabitation with the woman of
the realm of music not by himself singing but by listening, by appreciating what is sung to him, for him.

What the woman sings for him is that other featured song of the film,
Dixie Belle's "My Dreams are Gone With the Wind." In Chapter 1 I
described her performance here, posing as his sister, as her claiming to
have known him intimately forever, which is her way of constructing
their past together, a generic obligation. Her song and dance are meant
one way for Barbara Vance and her family and another way, a private
way, for Jerry. It is essential to her plot not merely that Jerry come to be
forced to leave this house but that he rejoice in his having a way out, or
anyway that he want to leave with her more than he wants anything
else. Her solution is to create her identity so that the very thing that
repels the proper Vances is what attracts Jerry, that he has a hidden, im-

proper sister.* He is of course impressed by the sheer daring of Lucy's performance, by the fact of it as well as by its content. His intimate, lucid smiles of appreciation acknowledge her emotional virtuosity, and they redeem, that is, incorporate, his earlier, hieratic, exalted smile at her discomfiture dancing with Dan in the nightclub.

But she wants and gets more than a spiritual victory, or more than revenge. The suggestion of sexual depth between them, and so also of sexual problems, is registered in her displaying the incestuous basis of their past. But her therapeutic move, let me say, is to demonstrate that what is his sister in her is not her ladylike accomplishments, as for example her trained voice and her ability to dance; his sister in her is what she shares with Dixie Belle, her willingness to lend her talents and her training to the expression of what Dixie Belle expresses, her recognition of their capacity to incorporate those improprieties. Her incorporation of familiarity and eroticism redeems both. And I would like to say further that she thereby redeems the fact of incorporation itself, that we live off one another, that we are cannibals. Thus she uses her sophistication, her civilization, to break through civilization to its conditions.

So it is not, to my mind, too much to say that on the way Irene Dunne plays this song and dance the recognition of her and her fictive husband's mystery, hence of the mystery of the film, depends. It requires the perfect deployment of her self-containment, her amused but accepting attitude toward the necessity for complication, for the pleasures of civilization; one could say it requires her respect for the doubleness (at the least) of human consciousness, for the comedy of being human, neither angel nor beast, awkward as between heaven and earth. For it is essential to its effect that her performance remain outside the song and its routine but at the same time show her awareness of its inner worth, to show both her difference from and her solidarity with Dixie Belle

* In another Leo McCarey film, *Once upon a Honeymoon* (1942), another woman's (Ginger Rogers's) identity is established for Cary Grant by the fact that she is revealed as someone surprisingly (given our introduction to her) capable of burlesque dancing. Here I recommend Robin Wood's valuable essay, "Democracy and Shpontanuity: Leo McCarey and the Hollywood Tradition," *Film Comment*, Volume 12 Number 1, January–February 1976. In *Together Again* (Charles Vidor, 1944), a fascinating, sometimes brilliant, but unsuccessful member of the genre of remarriage, an eminently honorable Irene Dunne, mayor of a New England town (in Vermont, not Connecticut), is briefly, scandalously, mistaken for a stripper.

and Dixie Belle's performance. You might call this the redemption of vulgarity by commonness.

I have been putting these responses to the song and dance of incorporation as from its sister's side. Put from its Dixie Belle side Lucy is declaring herself, to Jerry alone of course, as the woman he strays from the house to keep company with. She proposes herself as a field on which he may weave passion and tenderness, so that he might desire where he loves; or she reminds him of this possibility by reminding him who she is. Her proposal of herself as this kind of object is at once an offer and a challenge. It is not certain that either of them is really up to it. But her daring proposal is irreversible, and his exit with her means that he is taking her up on it.

THIS CALLS FOR CONNECTICUT, a chance for perspective and reconciliation, emotional and intellectual, say poetic or say philosophical. And again there is a problem about getting there. In *Adam's Rib* (as in *The Lady Eve*) you don't see the pair traveling there; it seems a place that exists mostly through the ambiguously projected extensions of a home movie, that is, it exists for movies. In *Bringing Up Baby* the road to Connecticut is paved with accidents, feathers, a sheriff directly descended from Dogberry, and a stolen car. In *The Awful Truth* the road also requires the infringement of the law and the abandoning of the everyday car in which you began the journey. Evidently the world elsewhere has its own laws and its entry demands a new mode, or new vehicle, of transport.

On the road to Aunt Patsy's place in Connecticut, Jerry asks Lucy if he can use her car to drive himself back to town after he drops her off. They are stopped either for speeding or for playing the car radio too loud, anyway for some species of joyriding. During the ensuing discussion Lucy sees to it (by releasing the brakes of her car and letting it roll into a ditch) that her car is not fit for further use that night—for example, not for a return trip. She is thus recreating a version of the scene she described for Jerry (and us) as she returned home with Armand that first afternoon. They are miles from nowhere but, unlike that earlier night when Armand's car broke down a million miles from nowhere and they had to find an inn, she and Jerry attract the help of two motorcycle policemen. When each of the pair is then shown being given a

THE SAME AND DIFFERENT

Having invited him to Connecticut to think again, she prompts him to think by her all but open sexual arousal, under the bedsheet, over the threshold, as the minutes edge away.

ride the rest of the way to Aunt Patsy's on the handlebars of a motor-cycle, one realizes that these vehicles are no less mythological than, say, the motorcycles in Cocteau's *Orphée*. Continuing their ride through the Connecticut countryside Lucy continues her drunk act. Bouncing up and down on the handlebars she sets off an exciting siren. She encourages Jerry to have fun too but his bounce produces only a choric raspberry. They are being driven back into the land of childhood, in this moment through the region in which little boys are disdainful of little girls. If Jerry does not know, on internal grounds, that this is different from anything that *could* have happened between Lucy and Armand, then he doesn't deserve to be here. The awful truth is that the truth of such matters can only be known on internal grounds.

(Again, for the last time, you can take the presence of these police-men not as messengers who transport those brave enough to demand happiness across the border from dailiness to comic enchantment, but as lackeys of the rich who make themselves available for the private purposes of those who are irresponsible because they own the law. My question is whether one of these views is less mythological than the other. Each is a total view, hence each is capable of accounting for the other. My conviction is that our lives depend on neither of them, as they stand, winning out completely over the other.)

They arrive at their destination less than an hour before midnight is to end their marriage, as a cuckoo clock will show that has not one door but a pair of doors and a pair of skipping persons appearing out of them every quarter hour, instead of a cuckoo or two or in place of gargoyles and virgins and knights and a scythed figure of time as death. It is another mythological object, a cinematic object, producible only on film. As Aunt Patsy has provided the locale for their conclusion she will provide Lucy's costume for it, a silk nightgown that Lucy is shown to tuck and tie up somehow so that it fits her. Evidently she needs not only encouragement and authority but instruction and preparation of a kind that a woman is fitter than a man to give. This would be why her "Aunt" appears instead of her father—or rather why, when it is this woman, at her phase of the story we are unearthing, whose Aunt appears, it is Aunt Patsy (Cecil Cunningham) and not the woman's aunt of *Bringing Up Baby* (May Robson), who when her friend Major Applegate suggests she might be capable of emitting erotic signals resembling a leopard's, responds "Now don't be rude, Horace." Aunt Patsy would have accepted the compliment.

Here is what happens in Connecticut. Lucy feigns first surprise that Aunt Patsy is not at the cabin and then a vast fatigue that sends her bounding light-heartedly upstairs to bed. Jerry is quite aware that her expressions of surprise and of fatigue are put on. Is his apparent resignation a sign merely that since he can't leave anyway (there won't be a car until tomorrow) he might as well see where this will all lead, putting himself in her hands, even if somewhat skeptically? Does he by now realize that she was not drunk during her act at the Vances'? And does he realize that this would have no bearing at all on whether she meant her incorporation of the familial and the erotic by one another, though it would have a bearing on how clear her memory of it is? His resigning himself, skeptically, into her hands is a continuation or confimation of his taking her up on her exit from the Vances'. He does not see how the thing is to be managed, what the road is that will lead back to their life together, but after her recent performance who knows what she is capable of?

Mr. Smith is not present, but after adjusting Aunt Patsy's nightgown to herself Lucy notices a black cat on the bed. Apparently their remarriage is to be dogged by a different muse, or totem, from that of their original marriage, or an additional one. Lucy shoos it off. I take it for granted that the black cat is a traditional symbol for female sexuality. Then does Lucy shoo the cat away because of what it stands for or because it merely stands for it?—as if to say: no more symbols of marriage, the real thing is about to take over. A rattling door comes open and in the adjoining room we see, through the open doorway, Jerry dressed in a nightshirt lent him by the caretaker. (This is, as far as I know, the original time that Cary Grant's sophistication and the kind of attractiveness he exhibits are tested by the mild indignity of a quasi-feminine get-up. A year later, in *Bringing Up Baby*, Howard Hawks will take this possibility to one of its extremes.) The cuckoo clock strikes for the first time in our presence, in close-up, to show 11:15—the marriage has forty-five minutes left to run, that is, the divorce has forty-five minutes in which to be headed off. The two childlike figurines, somewhere between live figurines and automatons, perhaps like animated figures of celluloid, appear from adjacent doorways in the clock and in parallel skip mechanically a few steps out, then turn and skip back in, the two doors closing with the last chimes. The house clock seems to be modeled on a Swiss chalet, and for all we know it is a replica of the country house our pair are now in.

In their respective beds, in adjacent rooms, as if their lives were parallel, not touching, and the skipping they had done together now seems mechanical, each looks at his and at her side of the same rattling door, silently urging it to open again. It does, upon which the pair mumble things about this oddity to one another. Jerry gets out of bed to, it turns out, examine and close the door, Lucy stays in her bed. He is still not able to see how to carry himself across the threshold. The clock strikes 11:30 and the figurines duly appear to celebrate the fact. Back in bed, Jerry notices that the source of the current that is causing the door to rattle and to open is coming from his partly raised window. He thereupon throws caution to the winds and raises the window all the way. For some reason, though the door rattles mightily it does not open. Can it be that their hopes are really gone with the wind—that unlike Dixie Belle's wind ex nightclub machina, Jerry's wind ex studio machina is going to fail, like just so much air? Lucy has to help some more. We are shown that it is the black cat that is stopping the door from opening, at first by lying in front of it, then when Jerry turns up the wind, by pushing desperately against the door with an outstretched paw, in a human gesture of, I find, unending hilarity. That cat knows that its hopes for an undisturbed life are due any second to be gone with the wind. After another moment Lucy notices the cat and shoos it out of the way again, as she had shooed it off the bed. With cooperation now from both Jerry and Lucy the door opens once again, this time discovering Jerry down on all fours, presumably from having been looking through the keyhole, presumably to discover what the barrier is to his dreams' opening up. That this discovers him to want the door open, while Lucy is left hidden in bed, is only fair: the invitation, the possibility of renewal, has been fully extended in her song and dance. But how can renewal come about?—the perennial question of reformers and revolutionaries, of anyone who wants to start over, who wants another chance. Even in America, the land of the second chance, and of transcendentalist redeemers, the paradox inevitably arises: you cannot change the world (for example, a state of marriage) until the people in it change, and the people cannot change until the world changes. The way back to their marriage is the way forward, as if to a honeymoon even more mysterious than their first. Taking a leaf from Plato's *Parmenides* they discuss their human plight in some metaphysical dialogue, the longest stretch of philosophical dialogue among the films of remarriage, the amplest obedience to the demand of the genre for philosophical speculation, for the perception that remarriage, hence marriage, is, whatever else it is, an intellectual undertaking, in the present instance, an undertaking that concerns, whatever else it concerns, change.

The relevant dialogue of this final sequence I find impossible to remember accurately, and it deserves preserving:

(The door opens for the second time.)

JERRY: In half an hour we'll no longer be Mr. and Mrs.—Funny, isn't it?

LUCY: Yes, it's funny that everything's the way it is on account of the way you feel.

JERRY: Huh?

THE SAME AND DIFFERENT

LUCY: Well, I mean if you didn't feel the way you feel, things wouldn't be the way they are, would they?

JERRY: But things are the way you made them.

LUCY: Oh no. They're the way you think I made them. I didn't make them that way at all. Things are just the same as they always were, only you're just the same, too, so I guess things will never be the same again. Ah-h. Good night.

. . .

(The door has opened for the third and last time.)

LUCY: You're all confused, aren't you?

JERRY: Uh-huh. Aren't you?

LUCY: No.

JERRY: Well you should be, because you're wrong about things being different because they're not the same. Things are different, except in a different way. You're still the same, only I've been a fool. Well, I'm not now. So, as long as I'm different, don't you think things could be the same again? Only a little different.

LUCY: You mean that, Jerry? No more doubts?

(Jerry doesn't answer her in so many words but says he's worried about the darn lock, the one on the door. Taking a cue from her glance he props a chair under the knob of the door but then seems surprised to find that he's locked them together in the same room. She lies back laughing.)

What I had in mind in referring to Plato's *Parmenides* was such a passage as this:

PARMENIDES: Then, that which becomes older than itself, also becomes at the same time younger than itself, if it is to have something to become older than.

ARISTOTLES: What do you mean?

PARMENIDES: I mean this.—A thing does not need to become different from another thing which is already different; it *is* different, and if its different has become, it has become different; if its different will be, it will be different; but of that which is becoming different, there cannot have been, or be about to be, or yet be, a different—the only different possible is one which is becoming.

ARISTOTLES: That is inevitable.

Philosophy, which may begin in wonder (thus showing its relation to tragedy), may continue in argument (thus showing its kinship with comedy). Human thinking, falling upon itself in time, is not required of beings exempt from tragedy and comedy.

Having invited Jerry to Connecticut to think again, Lucy prompts him to think by her all but open sexual arousal, under the bedsheet, over the threshold, as the minutes edge away. ("All but": he's still got to make a move.) The beginnings of philosophy in sexual attraction is how Plato sees the matter in *The Symposium*. Having once mentioned this vision in connection with Godard's films,* I am moved to mention here that the image in *Breathless* in which the couple climb together under a bedsheet, which then moves in patterns too abstract to read but unmistakable in erotic significance, has a precedent in the quite fantastic line of abstract impressions Irene Dunne invests her covering bedsheet with to signal Lucy's mounting arousal accompanying the tides of philosophy.

These signals of desire, and I suppose anxiety, are picked up from the opening mysteries of Jerry's absence from both home and Florida. If his cause was genuinely unrequited desire, or some dissatisfaction that adventures can make good, Lucy's new creation of herself is giving him a chance to right the balance. The price he will have to pay is, in his turn, that of change as well; he requires a move that will leave him different and, therefore, not different (because otherwise what would he be different *from?*). He must come to stand to himself in, say, the relation that remarriage stands to marriage, succeeding himself. (Can human beings change? The humor, and the sadness, of remarriage comedies can be said to result from the fact that we have no good answer to that question.) I spoke of Jerry's having to change as a price he must pay to right the balance of his marriage. I think of it as the price I ended up with in calculating Jerry's motive for his absences as the establishing of the possibility of freedom in marriage: he is going to have to find this freedom through remarriage. What this turns out to mean, at the conclusion of *The Awful Truth*, is that he can no longer regard a sexual imbalance in the marriage as the woman's fault.

I get there this way. I assume to begin with that there is a sexual brief each is holding against the other at the opening of what we know of their story. Jerry's "what wives don't know" mystery suggests this right off; it is confirmed by their never touching one another, after their

---

* *The World Viewed*, p. 100.

homecoming embrace is interrupted (except, as mentioned, at the end of the sister routine, and then as a brief theatrical walkaway); then their shared difficulties at the end with the door fit in with this line of thought. I assume further that Lucy's sister routine is not only triggered by Jerry's having made up the explanation or excuse for Barbara Vance's benefit that the woman in his apartment is his sister, but that the routine constitutes an answer to that explanation or excuse, a prophetic realization of it. These assumptions add up for me as follows.

Lucy's routine takes up Jerry's casting her as his sister as if it had been an explanation or excuse for *her* benefit, a statement of the cause of their loss of faith, of their faith in faithfulness, a loss in their sexual conversation. Then her song and dance for him that puts together kinship and desire is her reply to this excuse. I might translate her reply in something like these terms: Very well, I see the point. We do have this problem of having known one another forever, from the first, of being the first to show one another what equality and reciprocity might be. If this means being brother and sister, that cannot, to that extent, be bad. What is necessary now is not to estrange ourselves but to recognize, without denying our natural intimacy, that we are also strangers, separate, different; to keep our incestuousness symbolic, tropic, so that it joins us, not letting it lapse into literality, which will enjoin us. I'll show you that to be your sister, thus understood, will be to be stranger to you than you have yet known me to be. I am changed before your eyes, different so to speak from myself, hence not different. To see this you will have correspondingly to suffer metamorphosis. There is a wind effect here but you will just have to use your imagination.

So she gives rise to herself, recreates herself; and, it can be said, creates herself in his image, though it is an image he did not know he had or know was possible in this form. "The trouble with most marriages," Jerry announces in the second sequence of the film, preparing his sentences about faith, "is that people are always imagining things." It turns out that what is wrong is not with imagination as such but with the way most people use their imaginations, running it mechanically along ruts of suspicion. This causes, at best, farce, the negation of faith.

"You're all confused, aren't you?" she asks him, inviting him to work through the philosophy for himself. "Uh-huh. Aren't you?" His honesty deserves one further invitation, one last chance. "No," she offers him. It is the explicitness he needed. He was confused because he felt she was confused and he felt impotent to provide clarity for them. But if after all she is clear, that is another story. He casts his confusion about changing, becoming different, into words, thus making himself vulnerable to the therapy of love.

It is midnight. The figurine children skip out in their parallel paths to celebrate this hour of comedy. After they turn to skip back the boy is drawn to an escapement from the mechanism of time and accompanies the girl into her side of the habitation. The wind, an action of nature, that effects the closing of the door of marriage, is the work of no machine. We will have to imagine it for ourselves.

We are asked by this ending to imagine specifically how what we are shown adds up to the state of forgiveness the pair have achieved. In Connecticut the road back is to be found from what Jerry had called the road to Reno, which he characterizes as paved with suspicions. In *The Lady Eve* and in *Bringing Up Baby* and in *Adam's Rib*, as said, the discovery of the road back from divorce is explicitly entitled forgiveness; in *His Girl Friday* the place of forgiveness is taken by what the film calls a reprieve. Tracy's way of accepting George's suggestion in *The Philadelphia Story* that they "let bygones be bygones" is an acceptance of an interpretation of forgiveness as putting the past into the past and clearing the future for a new start, from the same or from a different starting place. I have at various junctures characterized this forgiving, the condition of remarriage, as the forgoing of revenge. When Tracy forgoes revenge toward George she finds nothing left for him. In Chapter 1 I took the experience of the end of a romantic comedy as a matter of a kind of forgetting, one that requires the passage, as it were, from one world (of imagination) to another, as from dreaming to waking, something that suggests itself as a natural way to describe the recovery from the viewing of a film as such. My adducing of *A Midsummer Night's Dream* in thinking about *The Philadelphia Story* offers an example of what this forgetting can look like. Emerson and Thoreau call the passage to this experience, I take it, dawn. The winning of a new beginning, a new creation, an innocence, by changes that effect or constitute the overcoming of revenge, extends a concatenation of ideas from Nietzsche's *Zarathustra*. In a related moment in "On Makavejev On Bergman" I quote the following from the section "Three Metamorphoses": "I name you three metamorphoses of the spirit: how the spirit shall become a camel, and the camel a lion, and the lion at last a child. There are many heavy things for the spirit, for the strong weight-bearing spirit in which dwells respect and awe: its strength longs for the heavy, for the heaviest ... To create freedom for itself and a sacred No even to duty: the lion is needed for that, my brother ... Why must the preying lion still become a child? The child is innocence and forgetfulness, a new beginning, a sport, a self-propelling wheel, a first motion, a sacred Yes." Camels of heavy marriages we know; and lions who can disdain them. A comic No to marriage is farce. I am taking our films to be proposing a comic Yes.

Nietzsche's vision of becoming a child and overcoming revenge is tied up with the achievement of a new vision of time, or a new stance toward it, an acceptance of Eternal Recurrence. And here we are, at the concluding image of *The Awful Truth*, watching two childlike figures returning, and meant to return as long as they exist, into a clock-house, a home of time, to inhabit time anew. How can my linking of Friedrich Nietzsche and Leo McCarey not be chance? How *can* it be chance?

All you need to accept in order to accept the connection are two propositions: that Nietzsche and McCarey are each originals, or anyway that each works on native ground, within which each knows and can mean what he does; and that there are certain truths to these matters which discover where the concepts come together of time and of childhood and of forgiveness and of overcoming revenge and of an acceptance of the repetitive needs of the body and the soul—of one's motions and one's motives, one's ecstacies and routines, one's sexuality and one's loves—as the truths of oneself. They will, whatever we discover, be awful truths, since otherwise why do truths about ourselves take such pains to find and to say?

On the way to these closed doors of marriage we have been given a moment that I recur to in my experience as to an epitome of the life of marriage that the films of our genre ask us to imagine, an image I take as epitomizing their aspiration to what I called a while ago life as festival, not something at the conclusion of a comedy but something of its character from beginning to end. I have in mind the conclusion of the sequence of the musicale (sequence 10), in which Jerry goes to interfere with an assignation and finds himself in the midst of a decorous recital. We know enough by this time of the practice of this kind of film to consider the sudden discovery of Lucy in front of the piano as the door flings open not as the surprising revelation that she is not after all engaged in an erotic form of life but that after all she is. Then it is her singing (whatever that is) that has been primarily felt by Jerry to be something beyond him, out of his control; not her singing teacher, who (whatever he is) is patently a secondary fiddle. Jerry, at any rate, is knocked to the ground by her performance here. His aplomb everywhere else is perfect. Lucy's strategy in her sister routine will require that he make the connection between her publicly singing a proper recital piece in a ladylike manner and her privately singing an improper piece in its appropriate manner. The epitome I say we are given of the life of marriage behind doors, for us to imagine, of marriage as romance, as adventure—of the dailiness of life, its diurnal repetitiveness, as its own possibility of festivity—is the moment of Lucy's response to Jerry's discomfiture as he tries to make himself inconspicuous at the unanticipated recital and winds up on the floor in a tableau with chair, table, and lamp. The spectacle he makes of himself starts a laugh in her which she cannot hold back until after she finishes her song but which pushes into her song to finish with it, its closing cadence turning to laughter. The moment of laughter and song becoming one another is the voice in which I imagine the conversation of marriage aspired to in these comedies to be conducted. We heard Lucy speaking to Aunt Patsy of the grand laughs she and Jerry have had. (All she will tell him, or warn him of, visiting him at his apartment, before becoming his sister, is that his ancient poem to her, which she is about to recite, will hand him a laugh.) At the musicale we are privileged to witness one of the grand laughs. This princess is evidently neither unwilling nor unable to laugh, indeed she generally seems on the brink of laughing. The truth is that only this man can bring her laughter on, even if he is sometimes reduced to poking her ribs with a pencil. This may not be worth half a father's kingdom, but she finds it, since he asks, worth giving herself for.

APPENDIX

FILM IN THE
UNIVERSITY

S there an honorable objection to the serious, humanistic study of
film? The idea of humanities is, grammatically and institutionally,
fixed in the plural. Thoreau is the only writer I know who uses the sin-
gular of "humanities" to refer to a kind of knowledge. His speaking of *a
humanity* is a part of his characteristic exaggeration, implying not
merely that the humanities are one with one another, as the sciences are
parts of science, but furthermore that they are not even one branching
of knowledge separate from the branching of the sciences, and indeed
not even separate from divinity (or from what at any rate deserves to be
called divinity). For normal people, however, who bitterly accede to the
plurality of humanities, what position is there from which we should
credit the suggestion that film is not one of their proper objects, or subjects? Was it not promised us that one day we might each become accomplished in any branch of labor we wished; to do one thing today
and another tomorrow, to hunt in the morning, fish in the afternoon,
raise cattle in the evening, and criticize after dinner? A university is
perhaps not the best place for all these activities, but then these were
surely only meant parabolically. Isn't a university the place in our culture that enables us now to teach one thing today and learn another tomorrow, to hunt for time to write in the morning, fish for a free projec-
tor in the afternoon, try to raise money for projects in the evening, and
after a seminar read criticism? To some this will not seem a Utopian set
of activities, but in the meantime, and for those with a taste for this particular disunity, why not have it?

The question whether film should be taught and studied seriously becomes less easy, and more edifying, when it is raised as a matter of curriculum; for then one is forced to consider what one is willing to pay for its study, which for many of us comes to considering what we are willing to forgo studying, and to have our students forgo studying, in order to study film. It invites us to make choices. For those of you for whom these choices present no special problems—perhaps because you have either from the beginning chosen film as a life's work or more recently decided upon a shift of career—the problems I have with these choices may seem beside the point. Then I ask the privilege of the stranger, to talk of his wanderings.

To orient my remarks about curriculum, I begin by aligning some quotations from two texts that I find repay study. I had to finish an earlier book about the topics I discuss here—*The World Viewed*—before I realized how much I was indebted to Robert Warshow's *The Immediate Experience* and how much my book might have gained had I known in time of Walter Benjamin and his essays and been able then to place myself in debt to them, especially to "The Work of Art in the Age of Mechanical Reproduction."

Both Benjamin and Warshow locate film for investigation in relation to high art on the one hand and to the experience of mass society on the other; and they are in agreement at a point that bears reiteration, or rediscovery, namely that however necessary it is to ask whether, or to assert that, film is an art, it is futile to ask the questions, or make the assertion, apart from a prior question. Benjamin puts it as "the primary question—whether the very invention of photography [and hence of film] [has] not transformed the entire nature of art"; and this turns out, in Benjamin's affirmative answer, to signal the transformation of the human senses, of human sense experience as such. Warshow puts the prior question as a conditional assertion: "Really the movies are . . . still the bastard child of art, and if in the end they must be made legitimate, it will be a changed household of art that receives them."

It may at once be objected to this yoking of such different writers that their subjects are different. Benjamin is talking about film, film in general and in relation to "world history," where this is conceived as "the adjustment of reality to the masses and of the masses to reality," a process, he says, of unlimited scope. Whereas Warshow is speaking of movies, movies mostly individually, and in relation, as he insists again and again, to himself, to *his* experience, to *his* reality. But this objection would conceal still bigger, and more surprising, connections.

It is true that Warshow and Benjamin face from opposite points of view the fact of "the increasing significance of the masses in contemporary life." Warshow formulates his problem by asking: "How shall we regain the use of our experience [that is, of any experience to call our own] in the world of mass culture?" Benjamin seems more confident, and would perhaps have noted Warshow's question as "unprogressive" or "superficial." For him, "The mass is a matrix from which all traditional behavior towards works of art issues in a new form"; "the greatly increased mass of participants has produced a change in the mode of participation." It is, he says, a fact that "the new form of participation first appeared in a disreputable form" ("It is a commonplace that the masses seek distraction whereas art demands concentration from the spectator"); but we must not be confused by this commonplace and this disreputable appearance: this very distraction, correctly analyzed, is a sign that film is being appropriated in the way art in all fields is now appropriated, a way that will allow film to provide "what one is entitled to ask from a work of art"—"the sight of immediate reality." The medium of film will answer this demand because the very material conditions of works within it—in particular, their inherent reproducibility (but really this should be seen as the inapplicability to them of the concept of reproduction; they are duplicated)—emancipates them from the value traditional art has placed upon the physical uniqueness of its works, upon what he calls their "aura." Hence film "exposes," hence depreciates, even shatters, the ritual and cult values within which works of art have previously been enshrined. Film provides a position from which to test, to examine, *present* historical occurrences. Benjamin finds in the provision of this position film's hidden "political significance," a significance that fascism must attempt to baffle and communism to organize.

But Benjamin's confidence and Warshow's anguish are themselves less signs of a difference in their grasp of film and of art than of differences in the historical circumstances in which they were writing. One such circumstance is comparatively universal: Benjamin, at the time of the writing I have quoted, had not had to face the issues named by the term Stalinism, whereas Warshow uses this term to name the experience he finds, "for those who were affected by it . . . [to be] the most important of our time; it is for us what the First World War and the experience of expatriation were for an earlier generation." A second such circumstance is comparatively individual: Warshow, a certain representative of American culture, puts forth the names of Henry James and T. S. Eliot as if staking his claim to a relation with high art; whereas the height of Benjamin's culture, a certain representative of Weimar, is so apparent as to go without saying. Their connection with one another, and their encouragement for me, lies in three of the lines of allegiance I find them to share: First, both refuse to exempt themselves from the mass response elicited by film (by some films, by some good films). Second, both claim that film poses a special problem, a revolutionary problem, for criticism and for aesthetics; they even specify their own contributions to this problem in what I take to be the same way: Warshow speaks of discovering in his work "a vocabulary"; Benjamin speaks of "[introducing . . . certain concepts] into the theory of art." Warshow seeks words that "begin to be adequate to the complexities of the subject, doing some justice to the claims both of art and of 'popular culture,' and remaining . . . in touch with the basic relation of spectator and object"; Benjamin's concepts are meant to be "completely useless for the purposes of Fascism . . . [but] useful for the formulation of revolutionary demands in the politics of art." Third, both intend their words in service of something that Warshow calls "the legitimization of the movies." Benjamin cannot allow himself to speak in such a way, since only a progressive, collective response will provide legitimization. But his efforts are directed, as he says Marx's were, so as to provide them with "prognostic value," and Benjamin's prognosis is of film's legitimization.

I am thus brought to the first of the questions that were proposed for the opening day of the conference for which these remarks were put together: "Should the study of film occupy a central place in a liberal arts or undergraduate curriculum?," and especially to what I understand as a recasting of this implied doubt in the opening assertion of the topics for the second day: "The study of film is only beginning to win a place for itself as a legitimate subject of scholarship."

I emphasize that I really wish to speak here of the prospect of contributing to a curriculum, not of teaching as such, which is a different, if not separate, matter. By the prospect of a curriculum, I mean the prospect of a community of teachers and students committed to a path of studies toward some mutually comprehensible and valuable goal; the goal will be subject to redefinition, but only by the methods of orderly and rational discourse through which the path to the goal is itself traversed. Whatever else the university is or has become, what makes an institution a university is its commitment to the idea of a curriculum, however much its practices fail to achieve it. No one will suppose that everything called teaching and learning film can, or should, happen only within an academic curriculum; any more than one should claim that everything that happens in a curriculum deserves to be called teaching. The beauty of a curriculum is that it can work (that is, something can be learned in it) in the relative absence of teaching. We know well enough its kinds of ugliness. The ugliness from which we run the greatest danger is the university's tendency to enshrine its subjects, to submit, or resubmit, the objects of its study to a kind of cult—ruled from what Nietzsche dismally describes as "The Chairs of Virtue"—something that is as hard to arise from as any cults in which those objects were created.

So aware have I been forced to become of the failings of the life of curriculum that I looked up some lines from a teacher for whom the beauty of a curriculum remained a perennial source of inspiration. Here are two sentences taken from the Introduction to *English Literature in Our Time and the University*, by F. R. Leavis: "Possibly most of those concerned about the issues know that the foundation at Cambridge towards the close of the first world war of the English Tripos was an important event in history. Nevertheless the contemptuous resistance offered to the idea of making, at an ancient English university, the critical study of English Literature the rival of Classics as 'humane education' is becoming difficult to recall." I have been through a number of curricular battles from the first year I started teaching for a living, each of which produced, in its time, contemptuous and, as academic affairs go, violent resistance; in most cases the violence and contempt are difficult to recall. Perhaps for that reason I do not so easily any longer find my faiths blind enough to see such changes in universities as historic events. So I find it cleansing to be made ashamed at having forgotten for a moment that curricular matters may be, should sometimes be felt as, matters of life and death for a teacher.

Leavis's faith in English literature, and in teaching, leads to two sobering points that guide me in thinking about what a curriculum is. (1) Given the intention and the opportunity of constructing a course of study, the realization of the best course of study is to be discovered only in practice, which means only in the experience of the particular people, and in the particular places, it is to occupy. Here I am agreeing with something Leavis says. I add that since a proposal for a curriculum is necessarily of prognostic value only, the more faith one has in the proposal, the more one's tone is apt to become positively prophetic. (2) My second guiding point is this. Given teachers with something to love and something to say and a talent for communicating both, you can afford to forget for a moment about curriculum. Whatever such teachers say is an education. And there are books the reading of which is also an education.

These two points—or assertions—provide the terms of a question I have about the study of film that I hope you will find honorable. It is the form in which the question of film's legitimacy, to the extent that I understand and share the question, presents itself to me. Because I know that the books whose reading I teach are better than anything I say about them; and because I believe that it is one, perhaps after all the fundamental, value of a teacher to put such books before students and to show that an adult human being takes them with whatever seriousness is at his, or her disposal; and because I know, furthermore, that the gift for teaching is as rare as any other human gift; my question is this: Is film worth teaching badly? And this is meant to ask: Does one believe that there are films the viewing of which is itself an education? I find that I have no stable answer to this question. (It is my version of a question I think is sometimes put this way: Does one mean by calling a film a masterpiece what one means in calling works in the established arts masterpieces? I do not suggest that the answer to such a question will settle the question of film's legitimacy. Some will say that there are no masterpieces in film, or none to be wished, and that for that very reason film is emancipating. One might seem to derive this idea from Walter Benjamin's remarks about mechanical reproduction depreciating the "aura" of the work of art.)

I emphasize the matter of curriculum for another reason as well. While I present myself both as a professor and as an advocate of film, that is, as one committed at once to the idea of a curriculum and to the idea that film should enter such a curriculum, there is no standing curriculum I know of that I am happy to see it enter, one that I am sure will do justice to them both. This is why I approach the question of a film curriculum personally, as a question of a something that I feel I can contribute to.

I will give a pair of examples of why or how my work as a teacher of philosophy in a university has brought me to the reading of films.

First an abstract example. Two of the philosophical writers for whom I have wished to assume curricular responsibility are Wittgenstein and Heidegger. They have, I believe, a number of significant features in common. One feature shows them as descendants of Kant, namely, their continuous appreciation and interpretation of the threat of skepticism, the possibility that the world we see is not the world as it is, that the world is not humanly knowable, or sharable. A second feature arises out of the first, namely their fundamental preoccupation with the fact and the concept of everyday life and their demanding a *return* of human thinking—in Wittgenstein's case *back* to the ordinary, the life of one's language to be shared with others; in Heidegger's case, *away* from what has become of that life and hence of that language. If I put their common preoccupation as a nostalgia for the present, it is to link them with a dominant experience of Romanticism. The writings of Warshow and Benjamin I was citing are related as well by their finding in the fact of film a materialization, even an agent for the overcoming, of this nostalgia for the present. The phenomenon is under discussion in *The World Viewed* through my interpretation of the fact and the concept of viewing as a particular awareness of one's absence from something present.

But the material conditions of film would not provide access to this philosophical nostalgia unless these conditions were given significance in individual films. This takes me to my second example, this time concrete. It again proceeds from a point in Heidegger.

What Heidegger calls Being-in-the-World is the basic state of what he calls *Dasein* (which is what we call the human). In the third chapter of *Being and Time* (the chapter entitled "The Worldhood of the World") he makes Being-in-the-World first visible—as a phenomenon for his special analysis—by drawing out, in his way, the implications of our ability to carry on certain simple forms of work, using simple tools in an environment defined by those tools (he calls it a work-world). (This is not unlike the imagery in the opening sections of Wittgenstein's *Phil-*

osophical Investigations.) It is upon the disturbing or disruption of such carryings on—say by a tool's breaking or by finding something material missing—above all in the disturbing of the kind of perception or absorption that these activities require (something that is at once like attention and like inattention) that, according to Heidegger, a particular form of awareness is called forth. (This so far sounds like John Dewey; so far, and so isolated, it is like John Dewey. But wait.) What this supervening awareness turns out to be *of* is the worldhood of the world—or, slightly more accurately, it is an awareness that that prior absorption was already directed toward a totality with which, as Heidegger puts it, the world announces itself. By the time Heidegger characterizes the supervening awareness as a mode of sight that allows us to see the things of the world in what he calls their conspicuousness, their obtrusiveness, and their obstinacy, one may sense an affinity with some of the principal topics of film comedy, especially silent comedy.

In contributing to a film curriculum, I should like to work out the idea that the comic figure whose modes of perception best fit Heidegger's phenomenological account in this early passage of his work is Buster Keaton. It is in Keaton's silent absorption with things (not, say, in Chaplin's) that what is unattended to is the worldhood of the world announcing itself (in the form, for example, of entire armies retreating and advancing behind his just-turned back). I should like to work this out in contributing to a *philosophical* curriculum as well. It is not unlikely that a department of philosophy as well as a department of film studies would object to such a proposal. Then I should interpret their objection—apart from matters of personality—as a denial either of the legitimacy of studying film, or of the legitimacy of studying philosophy, or of the legitimacy of studying Heidegger, or all three.

A more respectable objection to this proposal, from my point of view, might be put this way: it represents just one more instance of using film as an *illustration* of some prior set of preoccupations rather than constituting an effort to study the medium in and for itself, to gather what it specifically has to teach. This should serve to warn that one might pick up the Heidegger-Keaton connection without allowing it to prompt a study either of Heidegger or of Keaton. But to know what the illustration illustrates is to know what makes Keaton Keaton, something that requires knowing what makes Chaplin Chaplin and Griffith Griffith and Brady Brady and film film; just as to know what makes

Heidegger Heidegger requires knowing what makes Nietzsche Nietzsche and Kant Kant and Hölderlin Hölderlin and philosophy philosophy. It is because I do not know that any single person knows, or could know, all of those things—and other things that knowing them entails, some obvious, some unpredictable—and because I am nevertheless convinced that they are parts of some eventual unified study, that I am interested not only in the discussion, but in the formation, of a curriculum in film. In the meantime, those of us who have such an interest are, in good faith, going to have to do the best we can.

My readings of individual films provide my evidence of good faith in this matter of illustration.

One final plea. Walter Benjamin notes that "cult value does not give way without resistance"; he prophesies that film will overcome this resistance. Not only has this not come about (or not yet), but film may be taken to show that cults will form in the absence of unique objects, that the claim of uniqueness is deeper than the fact of uniqueness, that the claim will be made in the absence of the fact. I have spoken of a university, with its commitment to rational discourse toward some public goal, as if it too is an agent of the destruction of cults; but I have also admitted its own propensity to cultism. And I have spoken as if, for example, Wittgenstein and Heidegger, and perhaps Thoreau and Nietzsche, were clear candidates for a university curriculum, yet I know that each of them is mainly the object of a cult. None of them is the common possession of our intellectual culture at large, let alone our public discourse. It is possible that nothing is such a possession, that nothing valuable and comprehensible to each of us is valuable and comprehensible to all. And it is possible that every idea of value, like every object of value, must still arise as the possession of a cult, and that one must accordingly hope that some are more benign and useful than others.

I am moved to these last speculations because several times in the past months different people whose love and touch for film I respect have explained their energetic but ambivalent attention to various theories of film by saying that one or another of these theories is "the only game in town." What this means may be true. It may also be an expression of sad acquiescence in the reign of cults. Some of its causes are obvious enough. One grows weary of oneself with only oneself for conversation; and one gets cranky as well as hoarse; and—who knows?—the others seem so sure, they may be right. But the worst is that isolation causes uncreativeness and parochialism more often than it makes for anything better. I do not have to claim that everything is possible in every period in order to plead this much for universities: that while they may suffer every failing of the institutions of which they partake they are unique among institutions in preserving the thought that nothing is the only game in town, or that if something is, then there are habitations outside the town where it is not. For that reason, before any other, they have, as they stand, if not my devotion, my loyalty.

ACKNOWLEDGMENTS

Three of the readings in this book, in slightly different versions, have been published previously. The earliest, that of *Bringing Up Baby* (Chapter 3), was presented at a film conference at the Graduate Center of the City University of New York organized by Marshall Cohen and Gerald Mast in July 1975; it appeared in *The Georgia Review*, Summer 1976. The reading of *The Lady Eve* (Chapter 1) was presented at the annual meeting of the American Philosophical Association, Eastern Division, in December 1978, in Washington, D.C., and appeared in *New Literary History*, Summer 1979. That of *It Happened One Night* (Chapter 2) was my contribution to a symposium at Emory University in October 1979 entitled Intellect and Imagination: The Limits and Presuppositions of Intellectual Inquiry; it appeared in *Daedalus*, Spring 1980. I am grateful to the respective publications for their permission to reprint. A Rockefeller Humanities Fellowship for 1979–1980, together with a sabbatical leave from Harvard, gave me the time in which to write the bulk of these pages.

Thoughts of remarriage as generating a genre of film began presenting themselves to me during a course of mine on film comedy given at Harvard's Carpenter Center for Visual Studies in 1974. In 1976 I gave a version of this course designed to test out those ideas as rigorously as I knew how. In 1978 William Rothman and I offered a course jointly that took off from the material I had developed about remarriage and related it to other genres in (primarily) the Hollywood constellation of genres and to other films in which the actors and directors worked who were mainly responsible for the comedy of remarriage. That course was part of an expanded humanistic curriculum in film studies at Harvard made possible by a Program Grant from the National Endowment for the Humanities running from 1976–77 through 1978–79. The present book is accordingly one direct beneficiary of the talent and the equipment brought together at Harvard's Carpenter Center for the Visual Arts by the National Endowment's grant, and, it should be said, by Harvard's full participation in the costs of accepting that grant, on the part both of the administration of Harvard's Faculty of Arts and Sciences and of the faculty and staff of the Carpenter Center. Before and after the profit to me in that explicitly joint course, countless exchanges between William Rothman and me over the years, about film and more or less related matters, have affected these pages, as have his lectures and writings, for example, on Hitchcock, on Hawks, on Griffiths, on von Sternberg, on the documentary.

I have become indebted in more recent years to conversations with Norton Batkin about philosophy and about the arts, and about the ways between. He and Marian Keane assisted Rothman and me in our course, and their contributions to the discussions of the films would often make themselves felt as I pushed my readings further. In addition, Marian Keane worked with me in selecting the frame-enlargements that appear here, and she and Robbie Murphy refined a system for reproducing them from the prints of the films. In the earlier courses I mentioned I was assisted by Nicholas Browne. Without the kind of faithful managing he provided of the complications in mounting a course tied to multiple weekly screenings participation in such a course would not have been feasible for me in those years. Nor would my participation in such a course have come about as and when it did apart from the invitation by Epi Wiese to offer one of the set of discussion courses on film she was organizing in the late sixties and early seventies under the aegis of Harvard's program in General Education. That course, and indeed that program, have been succeeded at Harvard by other administrative accommodations, but their value, as they were, remains a permanent part of many lives.

Jay Cantor and Arnold Davidson responded to each chapter of the book as it was being drafted or revised; Susan Wolf was exceptionally helpful in her comments on the chapter on *It Happened One Night*. But many other friends, acquaintances, and relations have left their mark on what follows, sometimes, I dare say, unaware of their influence, in the small and in the large of it. It is a pleasure to list the names of Joyce Antler, Gus Blaisdell, Cathleen Cohen Cavell, Rachel Cavell, Marshall

Cohen, Timothy Gould, Stephen Graubard, Karen Hanson, John Harbison, John Hollander, John Irwin, Ira Jaffe, George Kateb, Hélène Keyssar, Mary Mothersill, John McNees, Martha Nussbaum, Joseph Reed, Camille Smith, Eugene Smith, and Garrett Stewart.

I presented versions of three of these chapters as Patten Lecturer at Indiana University in March 1980. Other versions and other chapters have been given at Amherst College; Georgia State University; Kalamazoo College; Middlebury College; Millersville State College; the University of California, Berkeley; the University of California, Santa Barbara; the University of Chicago; the University of Georgia; the University of Michigan; the University of New Mexico; Wesleyan University; and Yale University.

It is important to me to convey how specific and permanent have been certain of the suggestions I have picked up in discussions of these films. They need not have come just from friends like Timothy Gould and Kristine Korsgaard who in a discussion of *The Lady Eve* described Eve on the train as stalling, or like Norton Batkin who noticed that Eve's father at a critical moment is dressed in a wizard's robe. They can have come from a large class in which a student whose name I may not have learned asked a question I am not likely to forget—like where the men in these films get the time to spend all day pursuing love; or what I make of Cary Grant's repeating in *His Girl Friday* a line of his from *The Awful Truth*, "So you two are getting married." Or perhaps they have come during the discussion period after a more formal presentation, as when Newton Stallknecht at Indiana asked me how I would formulate the happiness these comedies are driving at if not as the claim that the pair lived happily ever after; or as when Stephen Donadio at Middlebury asked whether I found that a woman's sudden, shocking leap from a window in *His Girl Friday* is foreshadowed in the heroine's early gag line about jumping out of a window a long time ago. Some suggestions, of course, are still filed away for a more favorable opportunity than has presented itself to me since the time I received them. One correspondent, for example, has suggested that one pay attention to the photographs hung on the wall of the restaurant in *His Girl Friday*; another has suggested that the vehicles in *It Happened One Night* form as neat a system of significance as I claim its foods do; others have suggested that I pay more explicit attention to names—for example, to the Lords of Philadelphia, or to the fact that in a film in which the husband

ACKNOWLEDGMENTS

is named King Westley, Clark Gable is treated to an opening ceremony in which he is called "the King," or that in *Adam's Rib* Amanda's name incorporates a man and an Adam. So I must hope that it has been amply obvious that I have not thought of my readings as being complete in being exhaustive, but rather as being whole in being autonomous, in achieving sufficient directedness, sufficient integrity and extent, to arrive at conclusions, but ones which are provisional, so that others are prompted to continue them. (The idea of exhaustiveness in criticism is taken a little further in my essay on *King Lear* in *Must We Mean What We Say?*, pp. 311–313.)

One of the best gratifications, and confirmations, in this work of reading is that discussion, well prompted, allows those who take an interest to make clear and notable contributions to it. This bears the promise of a communal enterprise.

Because the present book is an extension of the preoccupation of my earlier book *The World Viewed* and of its addendum, "More of *The World Viewed*," I want to repeat here the acknowledgment of my indebtedness to Michael Fried for his role in helping me find ways to talk about film that met my experience of it. I add a word of thanks to my other colleagues at Harvard, besides William Rothman, in film studies, to Robert Gardner, Alfred Guzzetti, and Vlada Petric, for the welcome and the help they have given this wanderer from other precincts. They may not admire as much as I do the specific objects I am meditating on in this book, but their knowledge and love of film are a source of inspiration to all of us who work in and around the Carpenter Center.

A word about my dedication, to my son Benjamin, not yet five years old as I write this. Along with the other hopes between us, he stands in my mind on the present occasion for those younger generations of students and friends of mine who did not grow up with such films as this book considers but who are alive to their value and who are prepared—it is heartening to see—to accept the labors as well as the pleasures of their inheritance.

S.C.

July 4, 1981

INDEX